# Towards an Ecology of Brain

# Towards an Ecology of Brain

**Roger Walsh, M.D., Ph.D.**
Department of Psychiatry & Human Behavior
University of California Medical School
Irvine, California

SP MEDICAL & SCIENTIFIC BOOKS
New York

ISBN 978-94-011-9838-7 ISBN 978-94-011-9836-3 (eBook)
DOI 10.1007/978-94011-9836-3


Softcover reprint of the hardcover 1st edition 1981


SPECTRUM PUBLICATIONS, INC.
175-20 Wexford Terrace, Jamaica, N.Y. 11432

**Library of Congress Cataloging in Publication Data**

Walsh, Roger N.
Towards an ecology of brain.

Includes index.
1. Brain. 2. Human ecology. 3. Environmentally induced diseases. 4. Neuropsychology—Philosophy.
I. Title.
QP376.W233 304'.2 80-21727

*It is with appreciation and love that I dedicate this book to Patricia, Nugent, and Penelope Walsh*

# Towards an Ecology of Brain

# Contents

# CHAPTER 1
# Introduction

Let us then consider, for a moment, the world as described by the physicist. It consists of a number of fundamental particles which...appear bound by certain natural laws which indicate the form of their relationship.

Now the physicist himself who describes all this, is in his own account, himself constructed of it. He is, in short, made of a conglomeration of the very particulars he describes, no more, no less, bound together by and obeying such general laws as he himself has managed to find and to record.

Thus we cannot escape the fact that the world we know is constructed in order (and thus in such a way as to be able) to see itself.

This is indeed amazing.

Not so much in view of what it sees, although this may appear fantastic enough, but in respect of the fact that it can see at all.

But in order to do so, evidently it must first cut itself up into at least one state which sees and at least one other state which is seen. In this severed and mutilated condition, whatever sees is only partially itself. We may take it that the world undoubtedly is itself (i.e., is indistinct from itself), but, in any attempt to see itself as an object, it must, equally undoubtedly, act so as to make itself distinct from, and therefore false to, itself. In this condition it will always partially elude itself.

(G. Spencer Brown, 1972 p.104-105)

This type of dualistic search and knowledge, wherein the universe is severed into subject versus object, has been the very cornerstone of Western science and philosophy. But dualism has inherent limitations, which have rocked the prototype of Western dualistic science, physics, to its very core (Capra, 1975, 1976; Wilber, 1977). Erwin Schroedinger (1969), founder of quantum mechanics, put it bluntly:

> Subject and object are only one. The barrier between them cannot be said to have broken down as a result of recent experience in the physical sciences, for this barrier does not exist in the first place.

His conclusion is drastic indeed; "These shortcomings can hardly be avoided except by abandoning dualism." (Schroedinger, 1964).

> Abandoning dualism is exactly what the new physics has done. Besides relinquishing the illusory division between subject and object, wave and particle, mind and body, mental and material, the new physics - with the brilliant help of Albert Einstein - abandoned the dualism of space and time, energy and matter, and even space and objects.
>
> (Wilber, 1977; p.39)

Likewise, we are beginning to appreciate the notion that a purely dichotomous approach has its limitations in the neurosciences also. It is now apparent that the brain, the organ clearly designed to know the universe (and hence itself), cannot be fully described independently of the universe it is designed to know. At all levels from the size and weight of the whole brain, through neuron architecture, microstructure and function, down to synaptic size and structure and molecular biology, the brain reflects the history and current status of the interaction of this portion of the universe with the remainder. The history of their interaction is engraved in neural pathways, neural structure, chemical reactions, and molecular configurations.

The recognition of this interaction leads naturally to an ecological approach to the study of the brain, and this shift corresponds to related developments in a number of other scientific areas. Recent developments in physics, and the emergence of new fields such as developmental psychobiology, sensory deprivation, general systems theory, and various branches of ecology, all point to the increasing recognition of the importance of adopting a holistic perspective which includes the context within which an object is studied. In the case of living organisms this context is the environment.

An ecology of the brain represents a nodal point at the intersection of a variety of research areas: ecology, environmental studies, esthetics, sensory psychophysics, social structure and organization, general systems theory, interpersonal behavior, perception, neurophysiology, neuroanatomy, neurochemistry, neuroendocrinology, learning theory, and neural plasticity. All are involved in a study of environment and brain.

Of particular relevance here is the recent surge of interest in ecology. More books have been published in this area in the last five years than in the preceding five decades. In part this is the result of the current awakening to some of the deleterious effects of pollution and dwindling resources. Physical and chemical pollutants have been increasingly recognized as exacting a toll on both physical and psychological well-being.

The extent of their impact has been only slowly recognized, and is not yet fully appreciated. Thus, although the effects may be severe, it is often difficult to identify the causative agents because of such factors as the long lag phase between ingestion and the appearance of pathology, the over-determination of outcome and the interaction between numerous factors. There is no longer any question, however, that we are already paying a huge price for our ignorance and unwillingness to regulate our physical environment.

The sensory environment is more subtle but no less awesome in its power to effect long-lasting changes in physiology and behavior. Common sense and everyday experience have long recognized this as a major determinant of behavior, but whether the sensory environment is capable of modifying the very structure and function of the brain itself remained a moot point argued by philosophers and then scientists over centuries. The evolution of this debate is outlined in Chapter 2, but it is worth noting here that only within the last 30 years has definitive evidence been available and the question laid to rest. It is now apparent that only a few days of specific environmental exposure may be sufficient to elicit detectable anatomical and chemical changes, while for identification of chemical effects, stimulation for only minutes may suffice.

The implications and questions arising from this recognition are many. Firstly, with regard to the effects themselves, what are their nature, durations, and parameters, and their interactions with other factors? Some of these interactions may mean that research in a whole variety of fields now needs to be reassessed from an ecological perspective. Thus, for example, many of the effects long attributed to "malnutrition" may well reflect sensory rather than caloric input, and it is questionable whether more than about 5% of the research papers in this area have adequately controlled for this fact (Levine and Weiner, 1976; Richardson, 1976; Senf, 1976; Walsh and Cummins, 1976b, 1978). We must also ask what neural damage accrues from suboptimal or harmful sensory environments. There is now little question that in humans reared in suboptimal sensory environments both intellectual and emotional complications arise, and animal research suggests neural analogues of these.

A related question concerns the design of optimal environments. Maslow (1971) coined the term "eupsychian" for psychologically optimal environments, but whether there are "euneural" environments, whether they correspond to the eupsychian, and whether they can be identified, designed, and implemented are open questions. There is already evidence from both animal and human studies that appropriate sensory environments may facilitate behavioral recovery from brain damage, and concomitant animal research suggests that these same environments may facilitate neural recovery. Whether the brain changes which occur following environmental stimulation form the neural basis of memory is another important question yet only one of the many which arise when we begin to think of an ecology of the brain.

Almost all conceivable types of stimuli could be mixed together to provide an environment for experimental study. These could range from the naturalistic to the completely artificial. Just how naturalistic any laboratory environment can really be is a moot point, but this book focuses on those environments where there is some overlap with natural stimuli, and only examines in passing such totally artificial inputs as electric shock, and handling. The major focus is on environments which have differing amounts of social and physical stimulation. Similarly, although an environment exerts its influence on the total brain-behavior-body system of the organism, we will confine ourselves largely to the brain.

The purpose of this book is to provide a comprehensive review of the area by both offering an introduction to this field and at the same time allowing an overview and integration from which general principles and implications can be extracted.

The history of thinking about environmental effects on the brain has ranged from philosophical through psychological to neurological, and from positions of total disbelief to those of unquestioned acceptance. This evolution is examined in the next chapter, and is followed by the delineation of a model of research on the ecology of the brain that offers a framework within which to set and conceptualize data and thinking. To date there has been very little conceptualization of sensory environments, and so Chapter 4 attempts to provide a categorization and conceptualization of environments and their major stimulus dimensions. Chapter 5 examines the body of research on brain differences between animals reared in complex and stimulus-deprived environments, while Chapter 6 examines the differences between social rearing and deprivation. Chapter 7 then examines the role of sensory environments in the induction and therapy of brain damage. Once effects have been identified, an obvious and important question then becomes, what are the mechanisms mediating the production of these effects? Chapter 8 examines the mechanisms hypothesized to date.

Chapter 9 examines some of the far-reaching implications of the parallels involving this area of neuroscience, modern physics, and some of the consciousness disciplines, both ancient and modern. The concluding chapter provides an overview and a glimpse at the future of this field.

# CHAPTER 2
# A History of the Changing Brain

> Here it may be asked whether the organs (of the brain) increase by exercise? This may certainly happen in the brain as well as in the muscles; nay it seems more probable, because the blood is carried in greater abundance to the parts which are excited, and nutrition is performed by the blood. In order however, to be able to answer this question positively, we ought to observe the same persons when exercised and when not exercised; or at least observe many people who are, and many others who are not exercised during all periods of life.
>
> (Spurzheim, 1815)

Philosophical speculation on the effects of experience on brain structure and function has been going on for at least 2,000 years. Scientific research in this field, however, began only decades ago, and its respectability among scientists has been only recently established.

The major bone of contention throughout history has been whether the brain is perceptibly changed as a result of sensory experience, or whether it is to all effects and purposes immutable. This debate has been waged at different levels and in different disciplines for centuries, beginning with philosophy, progressing through psychology and psychiatry, and presently stirring the neurosciences.

The philosophical level of this debate was epistemological, that is, how do we come to know, how do our concepts and knowledge arise, and what is the relation between experience and the organization of the mind and brain? The two major opposing positions on this matter are empiricism and rationalism, and debate has raged under their banners for centuries. The empiricist maintains the position that experience is the only source of knowledge. Furthermore, all complex ideas and concepts are held to be built up out of simple sensations, and in turn are reducible to these elements. Thus, after examining the properties of the underlying elements, the empiricist can apply simple additive rules to characterize the properties of complex configurations (Hobbes, 1650). Mind is a "tabula rasa" on which experience engraves its memory (Locke, 1690).

The rationalists, on the other hand, argue that "although all knowledge begins _with_ experience, it by no means follows that it all originates _from_ experience" (Kant, 1781). They contend that to understand the derivation of real knowledge, one must presuppose a framework of perceptual and cognitive relationships, existing over and above the raw sensory data.

Inherent in the mind and brain, this framework is then projected onto the world. The rationalist theory of knowledge thus claims that it is not nature that imposes its truths upon us, rather we who project laws onto nature. The truths of the rationalist are not revelations concerning the existence of a world beyond the senses but rather insights into our own makeup. These insights provide frameworks and forms which hold and organize the contents of our experience (Hilgard and Bower, 1975).

At the psychological level, investigation into the extent of environmental modification of behavior has been a major focus of research, often simplistically conceptualized as "nature versus nurture." Early in this century, comparative psychologists began attempting to differentiate the effects of maturation and practice on the perfection of behavior patterns such as pecking in young chicks, or swimming in young amphibians (Harrison, 1904; Carmichael, 1926). At the human level, Freud's psychoanalytic postulates led to intense speculation and debate (Stachey, 1964; Rothgeb, 1971; Brenner, 1974), but remarkably little in the way of experimental investigation (Levy, 1934, 1938; Hunt, 1976). The recent explosion of systematic experimentation can be traced largely to Hebb's (1949) theories of the effects of perceptual experience on subsequent cognitive development, and to the European ethologists led by Lorenz (1952). "Nature versus nurture" still occasionally rears its head, and a tendency remains to oversimplify conceptualization of the etiology of behavior. But its multiply-overdetermined nature, and the interactive role of environmental effects, are being increasingly recognized (Meier, 1968; Walsh and Cummins, 1976b, 1978; Walsh and Greenough, 1976b).

At the level of the brain itself, there has been considerably less debate about the effects of environment and experience, largely because until very recently most people thought that there were no effects. With few exceptions, the zeitgeist favored the concept of brain immutability. As evidence to the contrary has accumulated, however, this zeitgeist has swung, though with considerable reluctance, to a view of the brain as a plastic organ, continuously molding and adapting to its environment.

The first person to test this question empirically was Malacarne (1785), who trained one of each litter-mate pair of dogs and birds over a period of several years. Although he concluded that the cerebella exhibited detectable changes, these claims made little impression, and for more than a century no one began further investigation.

Early studies were limited largely to investigating anatomical changes in sensory systems following accidental or surgical deafferentation (destruction of sensory organs). Typical early reports claimed that histological changes occurred in the optic system following enucleation (unilateral eye removal) in birds (Perlia, 1889), and in the mamalian cortex following tarsorrhaphy (eyelid suturing) (Berger, 1900) and reduced auditory and visual cortex size and cell numbers in a blind, deaf-mute human (Donaldson, 1890, 1891).

In addition to deafferentation studies, there were a small number of attempts to correlate neuron chemical and volumetric changes with functional states such as physical exhaustion (Luxenburg, 1899; Faure and Soula, 1913). These older studies have been reviewed by Lice et al. (1950), and Gomirato (1954).

Unfortunately the early work is of little more than exploratory significance, since it usually suffered from a lack of adequate controls and quantitative sophistication. To this day only the grosser effects of optic nerve and lateral geniculate atrophy have been found by utilizing nonquantitative techniques. Thus in the thirties the first major experimental review of the subject concluded that there was no firm support for the opinion that function could affect neural structure, since detectable changes were confounded by artifact and were too small to be meaningful (Goodman, 1932).

The turning point in deafferentation studies came with the demonstration of localized transneuronal degeneration in the lateral geniculate following unilateral enucleation (LeGros Clark and Penman, 1933-34; Glees and LeGros Clark, 1941). Shortly thereafter biochemical studies demonstrated alterations in neuron protein and ribonucleic acid (RNA) content. The brain's responsiveness to gross disruptions of its sensory systems was now beyond question.

However, the question of whether the brain would prove responsive to more subtle forms of sensory change remained moot, and the weight of opinion and zeitgeist inertia favored the negative. Early interpretations of environmental complexity reflected this belief. Thus Krech et al. (1960) reasoned that greater sensory input would increase cortical activity, acetylcholine (ACh) release, and induction of acetylcholinesterase (AChE), thus predicting an environmental complexity-induced increase in the concentration of this enzyme in the cerebral cortex. These experiments were run under the assumption that anatomical brain changes would be too small for detection. To their surprise, the opposite occurred. Not until two years later, when brain weights were analyzed, did the answer to the enzyme paradox become apparent. The reduced AChE activity turned out to reflect an increase in cortical weight greater than the increase in total amount of enzyme such that the additional enzyme activity had been diluted in the expanded cortex (Rosenzweig et al., 1962).

The initial scientific reaction to these findings was one of incredulity, which was overcome only by continued reports of consistent results and subsequent independent confirmation (Cragg, 1974). Moreover, further studies demonstrated that the amount of stimulation necessary to elicit such effects was considerably smaller than had been imagined, and that in fact two hours of exposure per day for less than one week could demonstrate significant changes.

In general, however, it was still assumed that these effects would occur only in the young, developing brain, and over a decade passed before this piece of neuromythology was overturned (Reige, 1971; Cummins et al, 1973).

Indeed, the question now arises as to whether any stimulus at all may not alter the brain, although whether the effects will be detectable or lost forever in neural noise is another question. The proper description of the brain may necessitate a definition of its environment, a situation strangely reminiscent of both modern physics (Heisenberg, 1963; Capra, 1975, 1976; Chew, 1968) and Eastern mysticism (Capra, 1975; Needham, 1956; Suzuki, 1968), which see the universe as an interconnected web of relations whose parts are defined only through their connections to the whole. At the very least the necessity of an ecological perspective for the study of the brain has been established.

## Chapter 3

# Towards a Model of an Ecology of the Brain

For the most part the environment is only a potentiality until actualized by appropriate actions: it is not a fixed property that inevitably impinges upon individuals.

(Bandura, 1977; p. 195)

Though the potential environment is identical for all animals, the actual environment depends upon their behavior. Is the animal controlling the environment or is the environment controlling the animal? What we have here is a two way process in which the organism appears either as an object or an agent of control, depending on which side of the reciprocal process one chooses to examine.

(Bandura, 1977; p.196)

This chapter is an attempt to provide a model of brain ecology which is both historical and conceptual: historical in the sense that studies of ecology and environmental influences on brain seem to occur in an evolutionary fashion; conceptual in that this evolution readily lends itself to understanding within a theoretical model of environment-brain interaction. It is based on a stimulus-response model in which the stimulus represents the environment and the response represents the effects on the brain. These effects are considered to be mediated by a variety of psychological, physiological, and chemical mechanisms, and in turn to exert a reciprocal influence on the environment (Fig. 1). As we will see in Chapter 9, a two way causality model such as this is actually overly simplistic, but is useful here in providing a framework within which to organize our discussion.

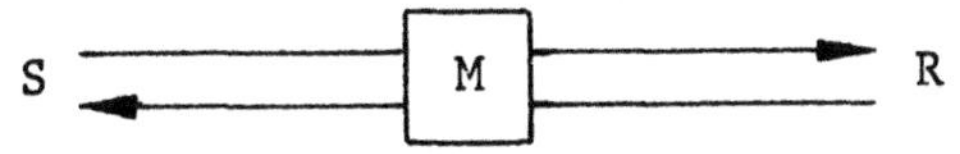


Fig. 1. A stimulus (S), response (R) model of the ecology of the brain. M=mediating mechanisms.

This model has five major divisions. The first of these is the response side of the paradigm: that is, just what are the qualitative and quantitative changes which occur in the brain as a result of environmental manipulation? Historically this was the first division to be investigated, and in fact the first demonstration of a brain response can be considered to represent the founding of this whole research area of brain ecology. Appropriately for a young field, most studies to date have been concerned with this division.

After responses have been demonstrated, interest tends to turn towards their temporal aspects. Research at this level tends to ask questions such as, what is the time course of the development, and sometimes regression, of these changes, and what effect does the age of the subject exert?

When a research field begins to feel more secure in its demonstration of a respectable range of responses, interest turns toward the stimulus side of the paradigm. The sensory stimulus employed to elicit demonstrable effects actually represents a multidimensional array of stimulus components, and the question is then posed, what are the precise stimulus attributes which elicit these effects? This stage is essentially one of component analysis.

In the fourth stage the effects of additional, independent variables are investigated. Whereas initial studies were concerned only with the effects of the sensory stimulation, more complex experimental designs are now employed wherein the interaction between this stimulation and other factors such as drugs, biological rhythms, and the genetic and experiental background of the subject are investigated.

The fifth level is concerned with the mechanisms which mediate the production of the observed findings. Ideally such a consideration should be a multi-level one, which takes into account mechanisms at a variety of levels, from the psychological through the physiological to the chemical. Since research tends to evolve in this fashion, the maturity of a field can be judged to some extent by the relative amounts of work accomplished at each stage. For brain ecology, the emphasis has been primarily on the first level, and perhaps least on the third (stimulus) and fifth (mechanisms). Each of these divisions will now be considered in greater detail.

## Responses

The brain responses to environment are most readily subdivided into anatomical, physiological, and chemical. As might be expected, the general trend has been one of moving from initial, fairly gross parameters to finer and more discrete ones. Thus, for example, some of the earliest anatomical measures were of cortical weight and size. When it was found that these gross measures did in fact respond, then the next logical question became, what are the histological components which contribute to these effects? Initial light microscopic studies tended to focus on such parameters as neuron, perikaryon, and nuclear size and neuroglial numbers. When these in turn were found to be reactive then smaller subcomponents such as dendritic length and branching were examined, and then dendritic spines, and finally at the electron-microscopic level, synaptic numbers and size. Thus the trend has been one of component analysis at finer and finer levels.

A similar pattern holds for physiological studies. Initial work tended to focus at the organismic level: for example, seizure thresholds and measures of drug lethality. Electroencephalogram studies emerged next and finally unit cell recordings in which the response characteristics of individual cells in various neural systems were assayed.

Chemically, the trend from gross to fine parameters is not quite as clear, but can perhaps be described as a shift toward less easily measured parameters (e.g., from cholinesterase concentration to catecholamine turnover assays).

A very large number of reactive parameters have now been identified and are examined in more detail in the following chapters. It seems possible that in fact all parameters may show some changes. The extreme viewpoint is that every stimulus will change, suggesting that negative findings may be due to a lack of experimental and measurement sensitivity, rather than to any lack of brain plasticity. We are thus moving towards a perspective diametrically opposed to the one held for so many centuries - that the brain is, to all intents and purposes, immutable.

Another trend is toward the detection of regional and parameter specificity. Regional specificity indicates that the magnitude of a response varies according to the brain region examined, as, for example, in the finding that the occipital cortex is the area most responsive to environmental complexity. Parameter specificity indicates that different parameters respond to the same stimulus with varying response magnitudes. Regional and parameter specificities may also interact such that one parameter may show a maximal effect in a certain area, whereas another will show its greatest response in a second region.

## Temporal Factors

The study of the temporal characteristics of the brain changes involves several dimensions. The first of these is usually the effects of varying the duration of exposure, and thus examines the time course of the evolution of these changes.

The second dimension looks at the effects of variations in the age at which treatment is initiated. Here the most commonly asked question is, is there a "critical" or "sensitive" period during which the brain is especially sensitive to stimulation? The most common assumption has been that the immature, developing brain will be the most sensitive. In some cases, monocular deprivation, for example, this has been found to be true: but for other forms of stimulation, like environmental complexity, this does not hold, and there is some evidence that the geriatric brain may be just as capable of response as the developing one.

The third area concerns the permanence and reversibility of environmentally induced effects, questions of obvious therapeutic importance. Here there are actually several subquestions, such as, will the effects be maintained after removal from the environment which elicited them? Can the extent of this maintenance be increased or decreased (reversed) by various means, including other environments? Does an optimum environment provide some protection against subsequent disadvantageous stimulation? And, is there a critical period within which treatment must be instituted to effect reversal?

The fact that it is possible to find interactions among the various temporal factors, and between them and other dimensions, complicates things. The effects of stimulation on a young animal may differ not only in magnitude but also in time course from those on an older one. Furthermore, different parameters may exhibit different temporal profiles.

## Stimulus

Research on the stimulus side of the S $\longleftarrow$ [M] $\longrightarrow$ R paradigm tends to be mainly a type of component analysis. Initial studies of environmental effects consider the environment as a whole, or a gestalt, and are not overly concerned with individual stimuli. After the environment has been demonstrated to be effective, however,the question arises as to precisely which stimuli are responsible for which effects. So while the initial approach is a unidimensional one in which the total stimulus impact along such lines as total input and novelty is considered(e.g., environmental complexity), later studies recognize the multidimensional nature of the stimulus complex.

In view of the multidimensional nature of the stimuli in any but the most barren environments, adequate description and conceptualization must necessarily entail description, preferably quantitative, of multiple dimensions and require a multidimensional grid. Development of this field of research, which is a form of progressive component analysis, tends to consist of movement from investigation of distal to proximal variables, from a unidimensional to an increasingly multidimensional conceptualization, and toward increasingly specific descriptions of the environment. To date there has been very little conceptualization of experimental environments, and so a possible system is described in the next chapter.

## Other Independent Variables

The fourth stage investigates independent variables other than the stimuli comprising the experimental environment. When these other variables are unrecognized or uncontrolled, they function as confounding variables which reduce the sensitivity, reliability, and validity of experimental findings. When they are recognized and appropriately manipulated and controlled for, they provide information on the interactions with environmental effects.

These variables can be divided into two major categories: those associated with the experimental subjects, and those independent of the subjects. The major subject variables can be grouped under the headings of genetic (species and strain), sex, age, and experiential background factors. Nonsubject variables are limitless in number, but common and significant ones include such things as drugs, hormones, and lesions.

## Mediating Mechanisms

The final stage comprises the study of the mechanisms which mediate the production of environmentally - induced brain changes. Obviously many mechanisms are involved at all levels, from the gross anatomical to the molecular biological, and it may be more appropriate to think of a hierarchy of processes than of a particular mediating mechanism. Once again the maturity of the field can be gauged to some extent by the amount of information available on multiple levels. Initially, research and theorizing tend to focus on only one or two levels, learning or arousal, for example - but as data and degree of sophistication increase, the implications of such mechanisms at both higher levels (organismic behavior), and lower levels, chemical processes, are tested. The provision of good theoretical models of mediating mechanisms is an important stage in the evolution of any area of research, since it provides a conceptual synthesis which allows the deduction and testing of multiple implications. Chapter 8 deals more fully with the topic of mediating mechanisms.

# CHAPTER 4
# Conceptualizing Sensory Environments

> The solution to the problem of infantile stimulation and subsequent behavior will only come when both independent and dependent variables are adequately specified.
>
> (Levine, 1962; p.155)

> Every organism...is in a constant process of adjustment to an environment external to itself. The life of an organism, in other words, is inescapably bound up with the conditions of the environment, which comprise other organisms and their activities as well...All organisms are engaged in activities which have as their logical conclusion adjustment to environment.
>
> (Hawley, 1950)

> Both historically and systematically psychology has forgotten that it is a science of organism - environment relationships, and has become a science of the organism.
>
> (Brunswick, 1957; p.6)

Although the rising tide of ecological interest has led to the study of ever-increasing numbers and varieties of environments, very little has been done toward conceptualizing and categorizing them. While some work has been done in this area for human environments, most notably by Rudolph Moos (1973, 1974, 1975, 1976), for animal studies there has been almost nothing. Researchers have instead tended to label environments in general ways (e.g., "social rearing"), and have only infrequently looked for the underlying commonalities and differences.

However, more sophisticated research demands more precise description, which in turn requires a system of conceptualization and categorization. The aim of this chapter is to examine the evolution, criteria, limitations, appropriate research strategies, and dimensions of such a system.

## Ways In Which Environment Modifies Brain

The environment may modify the brain directly or, when mediated by the sensory system or the internal milieu, indirectly. Direct influences may be physical, chemical, electrical, or radioactive. Such infuences are generally harmful, with rare exceptions such as those mentioned in the report of Bresler and Bitterman (1969), who found increased learning ability in fish whose brain size had been augmented by implantation. The brain's internal environment may also be modified directly by chemical or hormonal means, and each of these areas of intervention has its own extensive research literature. Of major interest to us are those effects wrought by sensory stimulation. The remaining discussion will focus on these.

## History And Evolution Of Conceptualization Of Environments

Since the initial stage of any environmental impact research is the simple demonstration of an effect, initial studies tend to use complex, multidimensional stimulus environments. At this stage neither independent nor dependent variables may be too clearly defined.

A logical next question asks: Which precise stimuli are responsible, and to what degree? The original unidimensional approach gives way to a more precise multi-dimensional analysis of stimulus components. At this stage, however, it not infrequently becomes apparent that different stimuli affect different response parameters to varying degrees, and so another order of complexity opens up and the precise research question becomes, which stimuli affect which responses? And finally, to complicate matters still further,there is a recognition that stimuli may exert interactive effects such that the influence of one stimulus is modified by another. At this stage a practical question may arise concerning the possibility of designing environments to exert an optimal influence of one kind or another. Thus, for example, the question arises as to the nature of a sensory environment which might elicit maximum recovery from brain trauma.

Along with this evolution in research questions is an accompanying evolution in research strategies and designs. Response detection gives way to component analysis, in which discrete stimuli or stimulus clusters which comprised the original environment are investigated individually.

The next phase, in which interactions between variables are investigated, necessitates more sophisticated experimental analyses, such as multiway analyses of variance or multivariate approaches such as linear regression analyses, factor analyses, and canonical correlation (Cooley and Lohnes, 1971). Such approaches indicate not only whether the stimulus has an effect but also the amount of variance for which it can account. In addition, they are productive of more sensitive, meaningful, valid, and ecological data than are traditional univariate approaches which fail to acknowledge the interactive, over-determined nature of outcomes (Senf, 1976; Walsh and Cummins, 1978). The general trend here is one of increased appreciation of the subtlety of environmental effects, along with a heightened awareness of their complexity and of the number of possible confounding variables. This in turn leads to greater sophistication in experimental controls and environmental manipulation. At this stage researchers sometimes recognize that the sensitivity of the brain is such that formerly ignored aspects of stimulation such as routine "maintenance" and animal house routines may have far-reaching effects on both brain and behavior. Thus Meier (1968), in discussing the conditions which modify development, stated:

> The conditions that have been described (e.g. handling, litter size, maternal strain and prepartal experience, cage size and complexity, to name a few), were previously relegated to the graceless limbo of maintenance routines which were infrequently recorded and codified, and even then for only limited or local consumption. Now we realize that these variables are awesome in their potency to effect behavioral change that is demonstrable for periods often extending far into adulthood. We cannot help but wonder how many of the now classic controversies in the behavioral sciences could not now be resolved with detailed information available on the maintenance routines employed by the protagonists in those controversies. (p.77)

Thus, as in many fields of science, what was yesterday's confounding variable becomes today's independent variable. In the area of environmental effects on brain and behavior it is probably dangerous to assume that any stimulus is incapable of causing detectable effects. This recognition of the brain's sensitivity, together with the evolution towards more sophisticated questioning, leads necessarily to a finer and more precise analysis of environment. Finer qualitative analysis leads to greater specificity and precision in the definition and description of stimulus dimensions and individual stimuli, including confounding variables. Thus, for example, maintenance routines and behavioral testing, traditionally noted in only a phrase or two, now must be more fully defined and described. Similarly, stimulus variables, such as the amount of lighting and number of aggressive interactions should be quantified wherever possible. This qualification is essential if one is to employ any of the more sophisticated experimental designs and analyses, such as those discussed above. Furthermore, at this stage of research it may become important to begin analysis of not only total stimulus input but also spatial and temporal patterns. Thus, for example, social interaction and toy exploration may show marked diurnal rhythms, and interventions at one phase may be significantly more effective than at others. Similarly, such measures as behavior density (the frequency of all types of activities at a particular place), behavior diffuseness (the range of different activities), and activity profile (the frequency of specific types of activity) all may be important determinants of social interaction effects. Such measures obtained for individual subjects begin to allow for analysis of environment - individual interaction and conceptualization of the environment - organism as an interrelated system.

Before examining the criteria and dimensions which might be used to conceptualize animal environments, let us take a look at those which have been used for humans, since they represent practically the only other systems to date. Moos (1973) lists six such dimensions.

1. Ecological dimensions, including meteorological and geographic variables as well as architecture and physical design. Both have been shown to exert a variety of behavioral influences, ranging from productivity to interpersonal behavior and organizational participation (Russet et al., 1964; Griffit and Veitch, 1971; Michelson, 1971). For reviews on the effects of man-made environments see Craik (1970) and Proshansky et al. (1970).

2. Dimensions of organization structure. This dimension emphasizes the role of structural dimensions such as size, staffing, salary, and span of organizational control. Reviews of various aspects of this view can be found in Pugh (1966), Roberts (1970), and Lichtman and Hunt (1971).

3. Personal characteristics of milieu inhabitants. This view implies that the character of an environment depends on the nature of its

members and their characteristics, such as age, sex, status, abilities, group memberships, physique, and other background data.

4. Behavior settings. These are seen as ecological units or niches and are conceptualized as being composed of both physical and behavioral components. This view originated with Robert Barker (1968), who emphasized the importance of studying behavior settings (classrooms, basketball games, etc.) as natural phenomena which represent behavior environment systems exerting significant coercive power over the behavior that occurs within them.

5. Functional or reinforcement properties of environments. This view emphasizes the role of reinforcement consequences in determining behavior, and is derived from behavior modification and social learning theories (Bandura, 1969, 1977; Mischel, 1968a, 1968b).

6. Psychosocial characteristics and organizational climate. This concept encompasses both psychological and social dimensions of the environment in a framework of person - milieu interaction. The approach accommodates both an introspective perception of what the environment is like and an outside observer's impression, although most of the work today emphasizes the importance of climate as perceived by participating members of the environment (Moos, 1973; Moos and Insel, 1974; Moos, 1976). Some of these categories, such as organizational structure and psychosocial characteristics, are distinctively human, but the others are at least partly relevant to animal studies, as we will see below.

The criteria by which dimensions of environmental categorization are chosen are rarely made explicit and in some cases probably go unrecognized. In the past there has been a certain arbitrariness, which is not surprising in view of the early state of the art. However, the criteria that can be identified appear to be of three main types. The first is a simple observation of the stimulus dimensions being used in current experiments. Thus, for example, the spate of studies on social rearing and agression readily point to significant categories. Second are the dimensions of demonstrated importance in other research fields. Thus, for example, the whole discipline of behavior modification point's to the importance of reinforcement. Thirdly, the direct observation of behavior is also suggestive.

## Limitations Of Dimensionality And Categorization

No conceptual system can ever fully encompass the reality it seeks to describe, and a conceptualization of environments is no exception. In view of environment's multidimensional nature any single dimension will of course cut across numerous others, and hence will be overlapping and nonexclusive. An adequate description of any environment may thus necessitate a multidimensional grid.

Any system exists in a state of continous flux, a point which is being made well by the general systems theorists (Miller, 1975). Slow change we label as structure, while we call faster changes process or function. Social environments are continuously in flux, and thus for precise studies it may be necessary to conceptualize a dynamic component which in turn may require temporal profiles.

Just as it is impossible to describe the organism independent of the environment, it may be similarly impossible to adequately describe the environment (which includes behavior) independent of the organism. "Any environment is a function of the conciousness which perceives it." (Ram Dass, 1975, p.9) Thus, it may be necessary to consider the basic unit an environment - organism system (Sargent, 1972) rather than a single environment entity.

## Dimensions Of Environmental Stimulation

In the present conceptualization outlined below 10 dimensions are identified.

1. Proximal - distal. This is the dimension suggested by Jessor and Richardson (1968), and essentially comprises a measure or indication of the directness of the stimulus effect on an organism or, alternately, of intermediary variables which mediate its effect. Proximal stimuli thus exert their influences relatively directly, whereas more distal ones influence the subject only indirectly, by acting via the mediation of more proximal stimuli. Thus proximal variables such as eye contact and biting might mediate the effects of more distal variable such as social hierarchy or pecking order. In general, most research fields are defined by distal variables, for example, social rearing and environmental complexity, but as research progresses the focus of attention shifts towards the proximal end.

2. Sensory modality. Here there are two major considerations: whether the stimulus acts through one or multiple modalities; and which modality (e.g., vision, or hearing). The sensory systems differ in the extent to which they lend themselves to environmental studies. To date the visual system has been most closely examined because of its relatively well-known neuroanatomy; its precise architecture whereby points in the visual field can be precisely located within its pathways; the wealth of related perceptual studies; and the precision with which at least some stimuli can be measured.

3. Social: Nonsocial. Social stimuli are of such importance to most mammals that a separate category seems clearly warranted. This is a distal variable which can be broken down into several levels of more proximal ones. Table I lists some of the more important.

TABLE I

SOCIAL STIMULI

- Characteristics of Members
  - Genetic background
    - species
    - strain
  - Sex
  - Age
  - Experiential background
- Number of Animals
  - Total number
  - Density per unit area
- Behavior of Others
  - Solitary, e.g. eat, drink
  - Interactional
    - play
    - aggression
    - dominance
    - sexual
    - nurturing
      - maternal
      - paternal

TABLE I (cont'd)

Behavioral Mapping

Behavior intensity

Behavior diffuseness

Activity profile

---

4. Afferent: Reafferent. Recently a cluster of studies has emerged demonstrating the importance of this dimension. Afferent stimuli are those which derive directly from the environment, whereas reafferent stimuli are at least partially generated by the organism itself (walking, watching oneself touch an object). It has been demonstrated that reafferent stimuli play an important role in both the development of and the recovery from brain damage of perceptual systems.

5. Intensity. This dimension refers to the intensity of stimulation.

6. Complexity. The complexity of an environment is a function of the range, variety, number, and dynamics of the stimuli.

7. Temporal. The time dimension has been a demarcation for a number of research areas. "Short-term" studies, minutes to hours, have been especially concerned with short-term biochemical processes, whereas "long-term" studies spanning a significant proportion of the life span have tended to include, in many cases, not only the biochemical but the structural (slow-changing processes [Miller, 1975]) effects as well. In addition, the timing and periodicity, of stimulation may be important.

8. Learned attributes of stimuli. The impact of a stimulus may be a function of an organism's prior exposure to it. Thus, for example, stimulus novelty elicits an arousal reaction, whereas repeated presentation tends to result in habituation. On the other hand, by virtue of their prior association with reinforcers, stimuli may acquire reinforcing properties and become secondary reinforcers. In this case they may become capable of eliciting markedly different responses than on first presentation.

9. Reinforcement. A reinforcer is defined as a stimulus which, when presented immediately after a behavior, is capable of modifying the probability of further emission of that behavior. Reinforcers are potent—some behaviorists would argue the major—modifiers of behavior.

10. The environment-organism system. Although this broad overlapping dimension has not yet been used in animal brain research, it is probably worth considering in view of the growing importance of general systems theory. The interested reader is referred elsewhere for reviews (Bertalanffy, 1969; Miller, 1975). This allows for the introduction of relevant concepts such as "intersystems congruence," which holds that the degree of congruence between interacting systems determines outcome. Thus the environment is viewed not as the determinant of behavior by itself, but rather as one system which interacts with others, such as biological rhythm and physiological state, to provide a setting in which specific behavior is more or less likely to occur (Michelson, 1970).

The above categorization is obviously far from complete and suffers from all the limitations previously discussed. However, it does provide a comprehensive system which includes the major dimensions relevant to current research. It is hoped that this system may have heuristic value in facilitating further research and conceptualization.

# CHAPTER 5
# Environmental Complexity and Isolation

> Does use and exertion of mental power gradually change the material structure of the brain, just as we see, for example, that much used muscles become stronger and that hard labor thickens the epidermis considerably? It is not improbable, although the scalpel cannot easily demonstrate this.
>
> (Soemmering, 1791)

## INTRODUCTION

The use of complex and deprived sensory environments was first begun by Hebb (1947, 1949), who enterprisingly and somewhat courageously reared rats as pets in his home. In comparison with those reared under impoverished laboratory conditions, his subsequently proved to be superior maze learners. The investigation of the effects of these environments on brain was first begun by the Berkeley research team composed especially of Mark Rosenzweig, Marian Diamond, Edward Bennett and David Krech, and any review of this area must read in large part as a tribute to the monolithic work of these people. Together they initiated a new field of careful, methodically replicated studies which have yielded a body of closely interwoven information rivaling that found in many more established fields.

Their initial interest lay in attempting to find correlations between problem solving ability and brain chemistry, in particular acetylcholinesterase activity (Rosenzweig et al., 1958, 1960). Having identified a correlation, they decided to test for the existence of effects of psychosocial and physical environments on this enzyme. They compared rats reared under conditions of environmental complexity with those reared in isolation, since these poles seemed to represent extremes of sensory input and were therefore expected to yield maximal differences.

Reasoning that greater input would in turn increase cerebral activity, acetylcholine release, neurotransmission, and induction of acetylcholinesterase, they predicted complexity-induced increases in this enzyme. To their surprise they found the opposite (Krech et al., 1960), with a reduced concentration in the cortex. It is fascinating to realize that these experiments, reflecting the zeitgeist of the times, were working under the assumption that anatomical brain changes would be to small for detection; it was not until two years later, when brain weights were analyzed, that the answer became apparent (Rosenzweig et al., 1962). In fact, the reduced cortical acetylcholinestrase activity reflected an increase in cortical weight greater than the increase in total amount of the enzyme. Further investigation revealed a regional response specificity and increased cortical depth as the basis of the weight change, and within the enlarged cortex the involvement of both neuronal and glial components. Recent studies have been concerned with changes at the cellular and electron microscope levels, temporal analyses, genetic, sex, hormonal, and drug effects and interactions, and the elucidation of mechanisms mediating the observed effects.

### Methods Of Differential Rearing

The differential rearing procedures employed by the Berkeley group will be described in some detail, since they have been used so widely and form a basis against which we can examine those used by other workers. Littermate male rat pups matched for body weight are assigned from weaning at about 25 days of age to one of two or sometimes three environments. Subjects exposed to environmental complexity (EC) are placed in groups of 10 to 12 in large cages (70 X 70 X 45 cms), to which a changing selection of stimulus objects, drawn from a pool of about 25, including such objects as ladders, wheels, light bulbs and chains, are added daily. These are diagrammed in Rosenzweig and Bennett (1969). In earlier studies animals were given 30 minutes of daily exposure to a Hebb-Williams maze in which the barriers were changed each day. In some experiments this maze experience involved actual training in problem solving for food reward (environmental complexity plus training), but this was later abandoned when it was found to add little to the cerebral effects of complexity alone (Rosenzweig, 1966).

Matched littermates of the EC animals are placed in smaller cages, either in isolation (isolated condition; IC) or in groups of three as a social condition (SC). SC and IC conditions employ cages of the same size (32 X 20 X 20 cms); but whereas the SC have wire mesh walls, isolates are usually placed in cages with solid metal walls, thus ensuring not only that they can neither see nor touch other rats, but also that external stimulation is minimal. More recently they have sometimes been placed in larger mesh cages similar to those of the SC group, but this has not been found to alter cerebral measures significantly. At the end of the differential rearing period, 30 to 80 days in the usual experiments, animals are killed and the subsequent anatomical and biochemical brain assays are performed blind: without knowledge of the experimental group from which they are derived.

## DEPENDENT VARIABLES IN BRAIN RESPONSES: NEUROANATOMY

### Whole Brain Weight

The effects of environmental complexity on whole brain weight are small and not to be sought without careful measurement of large numbers of matched subjects. Unless otherwise specified, the subjects mentioned in subsequent discussions are male rats differentially reared from weaning for 80 days, and the comparison is between EC and IC groups.

In the Berkeley labs the effect on whole brain amounted to a 1.0% increase significant only at the .01 level, in spite of their having run a total of 175 pairs (Rosenzweig et al., 1971). Using over 200 pairs differentially reared for periods from 18 to 530 days, Walsh et al. (1974) found a mean EC-IC difference of 2.6%. Significant findings have also been reported after 4 to 60 days by Ferchmin et al. (1970), and after 30 days (Geller et al., 1965). For the mouse, Henderson (1970) reported a 4.5% increase from birth, though this effect was not detectable in any of six parental inbred strains but only in their F1 and F2 crosses. However, LaTorre (1968) found increases of 5.7% and 4.5% in two inbred strains, and these were larger than

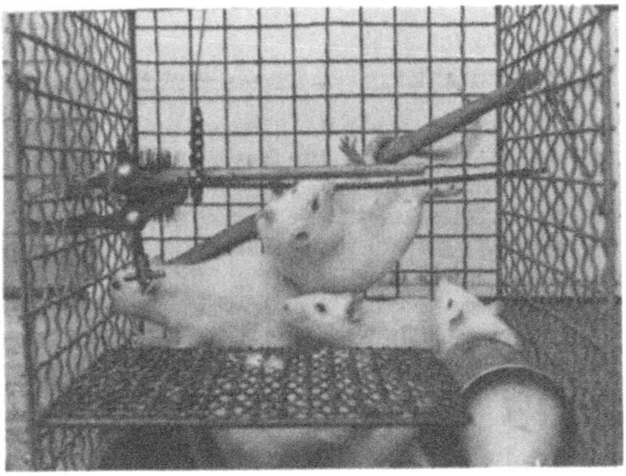

Fig. 1. The Environmental Complexity (EC) condition.

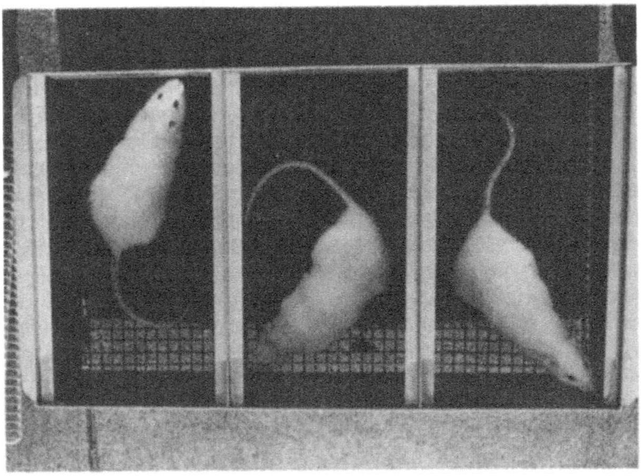

Fig. 2. The Isolation Control (IC) condition. Solid walls separate individual cages.

Fig. 3. Another perspective on differential rearing environments.

those most often reported for the rat. Not suprisingly, several workers using smaller numbers of subjects have been unsuccessful in finding significant effects.

The reasons for the small size of the whole brain response are to be found in the low variability of the brain and the regional localization of response. It should be noted that although for ease of description the effects are often attributed to complexity, this perspective is arbitrary, and they could be attributed, just as arbitrarily, to isolation. A more correct but lengthy statement would note that EC-IC differential rearing elicits differences and it is only these differences which can be measured.

## Regional Specificity

The increase in weight after exposure to environmental complexity is not uniform throughout the brain, but displays instead patterns of regional and temporal specificty. The terminology used to describe this regional localization bears only a partial relationship to the functional and anatomical natures of the areas, but rather it derives from and reflects the method of brain dissection evolved by the Berkeley workers. This is performed with the aid of a plastic T square, placed over the brain, whose markings guide a freehand dissection with which samples of occipital, somesthetic, and remaining dorsal cortex are peeled off from the underlying corpus callosum. Finally, all of the remaining cerebral tissue is stripped from the underlying subcortex and ventral cortex, which comprises not only ventral cortical tissue, but also contiguous organs including amygdaloid nuclei, hippocampus, dentate gyrus and corpus callosum (Rosenzweig et al., 1962). The remaining tissue, composed of olfactory bulbs, diencephalon, midbrain (mesencephalon), and hindbrain (rhombencephalon), is referred to as subcortex.

Within the cortex, effects after 80 days are maximal in the occipital region (6.4%), with the next most marked finding occurring in dorsal cortex (5.0%), then ventral (3.0%), and least (1.9%) in the somesthetic area. For the cortex as a whole (total cortex) this yields a mean effect of 4% which contrasts with a decrease of 1.1% in the remaining brain or subcortex (Rosenzweig et al., 1971). Since these cortical and subcortical effects are in opposite directions, the cortical subcortical ratio (C/S) yields a larger and more sensitive effect (5.2%) than would either alone (Zolman and Morimoto, 1962, 1965; Rosenzweig et al., 1971)(See Tables 2, 3 and 4). Being a within-subjects ratio, the C/S ratio largely removes effects of individual differences due to factors such as genetic background, nutritional status, and body weight and, hence, makes it a more sensitive measure of environmental effects. To what extent this ratio reflects patterning of neural activity within the brain is uncertain.

Regional localization has not been so precisely delineated in the subcortex. However, significant responsiveness may be limited to the forebrain midbrain areas (diencephalon and mesencephalon), since no effect was found on total hindbrain weight (Ferchmin et al., 1970; Walsh et al., 1974) or hindbrain components (Rosenzweig et al., 1962). A more discrete localization may exist, since a significant difference in width was found in the area underlying the frontal (Walsh et al., 1973b) but not the occipital (Diamond et al., 1966) cortex; and in the mesencephalon both the superior and inferior colliculi have been found to be heavier after complexity (Krech et al., 1963). It has been suggested frequently that extracellular fluid might be responsible for the observed weight changes. But since both wet and dry measures yield the same effects (Bennett et al., 1969; Eterovic and Ferchmin, 1974; Hoover and Diamond, 1976), we must conclude that any fluid changes are minimal.

## The Pineal

Though never empirically refuted, the pineal's original role as "the seat of the soul" has been largely disregarded, and it is now thought to function as a neuroendocrine transducer particularly involved in reproductive physiology. Since the discovery of a number of endocrine changes following complexity isolation rearing, e.g., pituitary, thyroid, adrenal, thymic, testicular, liver and splenic weights (Hatch et al., 1963; Walsh et al., 1971) it seemed possible that complexity might modify the pineal. However, the only published report to date (Quay et al., 1969) found that social density, but not complexity per se, altered pineal weight; but neither factor influenced mid-day 5 hydroxytryptamine concentration or acetylserotonin methyltranferase activity. Since the relative contributions of social and physical factors to the induction of the complexity isolation endocrine differences listed above are unknown, it is impossible to be sure of the role of the pineal in their production. Further studies which separate the effects of these factors will be necessary before an answer can be given.

## Cerebral Dimensions

The growth of the rat brain is not complete at birth nor at weaning, but in fact continues throughout life, though only very slightly after about 75 days. In the cerebrum, growth in width is virtually complete by 20 days, but length continues to increase beyond 90 (Bennett and Rosenzweig, 1968; Altman et al., 1968c; Walsh et al., 1971a, 1971b). This growth is susceptible to environmental stimulation, since four months of operant conditioning, ten days of handling during infancy, and three months of environmental complexity were all found to increase cerebral length but not width (Altman et al., 1968b). In contrast, however, Rosenzweig and Bennett (1969) were unable to detect significant dimensional effects on either rat or gerbil cerebrum after 30 days of differential rearing.

This apparent contradiction was resolved when Walsh et al. (1971a, 1973a) found that whereas 80 days differential rearing elicited a highly significant 2.5% increase in length, the 30-day effect was considerably smaller 1.2%. A subsequent series of experiments have confirmed the cerebral length increase for differential rearing periods ranging from 18 to 530 days (Cummins et al., 1972, 1973; Walsh et al., 1973b; Kuenzle and Knusel, 1974). Cerebral width effects have not been detected directly, but the product of length and width has proved the most powerful EC-IC discriminant. Perhaps environment exerts a similar though smaller effect on width as on length, but this effect is too small to be detected by itself and is made apparent only when a product of the two measures is obtained.

In their original observation of behaviorally-induced changes in length Altman et al. (1968c) drew attention to the presence of a mitotically active subependymal layer in the anterior forebrain of adult rats. Cells from this layer migrate to the anterior cortex and basal ganglia. Walsh et al. (1973a) therefore reasoned that if modification of cell division in this layer underlay dimensional changes, they would then be localized in the anterior regions of the cerebrum. This has now been found to be the case (Walsh et al., 1973a), and raises the interesting question of the relationship of the cerebral length changes to those of cortical weight and depth. Since the length effect is limited to the anterior region, and the latter are more marked posteriorally, they may well be largely independent of one
It is unknown whether the magnitude of the increase in length following complexity is a function of the unstimulated growth rate, but it would be fascinating to determine the effects of additional stimulation in the first 20 days of life, while growth is rapid and the skull sutures remain open.

## Cortical Depth

Following the repeated findings of an increase in cortical weight, the logical question was whether an underlying morphological substrate for this change could be found. Microscopy provided an affirmative answer, with the finding that changes in cortical depth paralleled those in weight (Diamond et al., 1964, 1966; Diamond, 1967; Walsh et al., 1972; Szeligo and Leblond, 1977). After 80 days of differential rearing from weaning, the cortex is deeper in EC than IC animals. In general the effect is greater in the medial than in the lateral cortex, the former being more susceptible to increases and the latter to decreases in environmental complexity. It is maximal in the medial occipital cortex (around 7%) and minimal in the motor somatosensory regions (Rosenzweig et al., 1971). However, in the medial portion of the frontal cortex an EC-IC difference of 5.5% was noted within an area 1.5 to 3.5 mms from the midline, being maximal medially (8.5%) (Walsh et al., 1972).

In summary, there is now considerable evidence that the depth of the rat cortex is modified by complexity - isolation in a regionally specific manner, but the functional significance of this patterning remains unknown. The cortical increase is clearly significant but not yet sufficiently large to fulfill Marian Diamond's long held dream of causing the rat cortex to form its first convolution. In general the correlations obtained between weight and depth measures have been good (Diamond et al., 1972). The advantages of depth as opposed to weight measures are that the former permit a much greater degree of spatial localization of changes than do the latter, and also preserve the brain for subsequent histological studies.

## Hippocampal Thickness

In view of its function in long-term memory formation, to anticipate hippocampal involvement following complexity would seem logical. To date, however, results have been largely negative, at least as regards size and thickness, perhaps due in part to the marked individual variability which this organ displays. Walsh et al. (1969) reported that in the medial region underlying the occipital cortex, the EC hippocampus was 5.7% thicker than that of the IC after 90 days, though this difference was significant only at the .05 level. However, measures by the Berkeley team of both weight and thickness, using several different ages and durations of differential rearing, have been unsuccessful (Diamond et al., 1976; Rosenzweig and Bennett, 1978) although findings of changes in neuroglia, neuron size, and dendritic morphology (Walsh et al., 1969; Cummins et al., 1974b; Walsh, 1975; Fiala et al., 1978) clearly indicate significant hippocampal plasticity in response to the environment. At the present time therefore, it must be concluded that if complexity affects hippocampal depth, the effect is small and obscured by wide individual differences.

## Corpus Callosum

As a consequence of the increased cortical activity following complexity, one might anticipate a greater transfer of information between hemispheres, and consequent functional - anatomic - biochemical modification of the major transfer organ, the corpus callosum. And indeed Rosenzweig (1970) has demonstrated augmented transcallosal electrical conductivity, and Altman and Das (1964) have indicated an increased number of autoradiographically labeled cells in the callosum from complex animals. More recently they have interpreted this finding as an indication of greater production by the ventricular subependymal layers of new neurons and glia which subsequently migrate via the corpus callosum and other pathways to their definitive cerebral positions (Altman, 1970; Altman et al., 1968c). The actual size of the corpus callosum may also be altered, since Walsh et al. (1972) found that in the frontal but not occipital region the lateral but not medial callosum was 13.6% ($p < 0.01$) thicker following complexity. It is difficult to interpret this finding, but if replicated it may indicate a cerebral region where complexity elicits a high degree of interhemispheric information transfer.

## Diencephalic Dimensions

It has been found repeatedly that while the weight of the cortex increases following exposure to environmental complexity, that the subcortex decreases (Rosenzweig et al., 1971), though to a relatively lesser extent. The microscopic substrate of increased cortical weight was found to be an increased depth, and therefore it seems logical to anticipate that microscopy might detect a reduction, albeit small, in the size of the diencephalon. Diamond et al. (1966) were unable to detect such an effect in the occipital region, while Das and Altman (1966) in the previously cited study reported an increase of 7% in the cross sectional area. Significance levels were not cited, and it should be noted that the trend is opposite to that anticipated from weight. Walsh et al. (1972) measured both breadth and depth of the diencephalon underlying the frontal cortex and found them to be reduced in EC by 3.8% and 3.3% respectively, although only the former reached significant levels. These findings are in the predicted direction and were obtained blind, with larger numbers of subjects (15 littermate pairs) than other studies. They therefore suggest the possibility of subcortical size changes analogous to the depth changes identified in the cortex. Further studies will be necessary to confirm them and to determine whether a detectable degree of subcortical regional specificity exists. The absence of any significant degree of localization for weight (Rosenzweig et al., 1962) suggests that dimensional specificity, if it can be identified, will be small.

Possible candidates for the detection of such changes might include the colliculi, in view of their observed weight changes (Krech et al., 1963), and the lateral geniculate because of the alteration in neuron nucleus: neuropil ratio (Malkasian and Diamond, 1971). On theoretical grounds one might also expect cerebellar components, since studies in related fields point to the necessity for active self-guided exploration and kinesthetic feedback (reafference), in addition to passive examination of the environment from a distance (exafference), in order for learning to occur. Furthermore, active interaction with the toys appears necessary for the full cerebral EC effect; and since the cerebellum is the major kinesthetic feedback and integrative organ, environmental effects may well be detectable. Prescott (1970) in particular has laid theoretical emphasis on the role of kinesthesia and the cerebellum in development of the central nervous system (CNS). However, since it is a complex heterogenous organ, there may well be a degree of response localization, such that while effects for the whole organ are extremely small (Rosenzweig et al., 1971) various components may be particularly affected. This is analogous to the small effect on the whole brain weight.

Table 1: Summary of Anatomical Changes Measured in the Medial Occipital Cortex. Based Partially on Diamond (1976)

| Measurement and Condition | N (pairs) | Experimental duration in days | Percent difference | P |
|---|---|---|---|---|
| I. Cortical depth difference between enriched and impoverished rats | | | | |
| A. Preweaned Long-Evans males | 7 | 6-14 | 2 | N.S. |
| | 7 | 6-19 | 7 | 0.01 |
| | 19 | 6-19 | 9 | 0.01 |
| B. Wistar males (Walsh et al., 1969) | 8 | 25-105 | 4.9 | 0.05 |
| C. Postweaned $S_1$ males | 24 | 60-61 | 0.7 | N.S. |
| | 20 | 26-30 | 4 | 0.001 |
| | 24 | 60-64 | 3 | 0.01 |
| | 22 | 25-32 | 4 | 0.001 |
| | 23 | 25-40 | 7 | 0.001 |
| | 30 | 25-55 | 7 | 0.001 |
| | 50 | 60-90 | 7 | 0.001 |
| | 22 | 25-105 | 8 | 0.001 |
| | 18 | 105-185 | 5 | 0.05 |
| | 22 | 25-185 | 3 | 0.05 |
| D. Castrated Long-Evans males neonates | 11 | 60-90 | 6 | 0.05 |
| E. Castrated Long-Evans males at 30 days of age | 12 | 60-90 | 7 | 0.001 |
| F. Nonpregnant Long-Evans females | 23 | 60-116 | 4 | 0.001 |
| G. Postpartum Long-Evans females | 21 | 60-116 | 3 | N.S. |
| H. Newborns from Long-Evans enriched and impoverished | 30 | 95-116 | 0 | N.S. |
| II. Cortical depth difference between enriched and standard colony postweaned $S_1$ males | 20 | 26-30 | 0.5 | N.S. |
| | 24 | 60-61 | 4 | 0.001 |
| | 23 | 25-40 | 3 | 0.05 |
| | 22 | 25-55 | 3 | 0.05 |
| | 21 | 60-90 | 5 | 0.001 |
| III. Cortical depth differences between standard colony and impoverished $S_1$ males | 20 | 26-30 | 3 | 0.01 |
| | 24 | 60-64 | 1 | N.S. |
| | 23 | 25-40 | 5 | 0.001 |
| | 22 | 25-55 | 2 | 0.05 |
| | 21 | 60-90 | 4 | 0.05 |
| IV. Difference in glial number betwen enriched and impoverished $S_1$ males | 17 | 25-105 | 14 | 0.01 |

Table I (continued)

| Measurement and Condition | N (pairs) | Experimental duration in days | Percent difference | P |
|---|---|---|---|---|
| EC-IC glial differences (IC and SC did not differ significantly) (Szeligo and Leblond, 1977) | | | | |
| Oligodendroglia | 14 | 20-50 | 27 | 0.01 |
| | 10 | 20-100 | 33 | 0.01 |
| Astroglia | 14 | 20-50 | | N.S. |
| | 10 | 20-100 | 13 | 0.01 |
| V. Difference in neuronal area between enriched and impoverished $S_1$ males | 12 | 60-90 | 4 | 0.05 |
| | 20 | 25-105 | 12 | 0.001 |
| VI. Difference in neuronal nuclear area between enriched and impoverished $S_1$ males | 20 | 25-55 | 2 | N.S. |
| | 19 | 60-90 | 2 | N.S. |
| | 23 | 25-105 | 12 | 0.001 |
| | 18 | 105-185 | -1 | N.S. |
| VII. Difference in post-synaptic length between Enriched and standard colony $S_1$ males | 13 | 25-55 | 1 | N.S. |
| Standard Colony and impoverished $S_1$ males | 13 | 25-55 | 6 | 0.05 |
| Enriched and impoverished $S_1$ males | 13 | 25-55 | 8 | 0.01 |

## NEUROHISTOLOGY

Having identified dimensional changes within the cortex as resulting from environmental complexity and isolation, we ask the next logical question: What constituents contribute to these changes? This question has been more methodically and carefully analyzed here than in related research areas such as early handling and dark rearing; and a variety of histological changes ranging from cell numbers, types, and sizes to synaptic numbers and sizes have been identified.

### Cortical Differential Cell Counts

#### *Neuron Density*

In the standard 80-day experiment, cell counts have revealed a decrease in neuron numbers per unit cortical volume in complex as compared with isolated animals. Diamond et al. (1964) found that in the occipital cortex the reduction per microscopic field amounted to 17%, was maximal in layers II and III, was not apparent in layer I, and reversed in layers IVa and Vb. In the somatosensory cortex the reduction in density was less marked (7%), and occurred in layers II, IV and V, while layers III and VI were less consistent and occasionally showed reversal. A similar pattern was noted by Cummins and Walsh (1978), who found a 13.5% effect which was not evident in layer I, was maximal in layers II, III and IVa, and was reversed in layers IVc and Va.

The absence of depth changes in layer I has been reported following direct measurement in several areas (Diamond et al., 1964, 1966; Walsh et al., 1969, 1972; Cummins and Walsh, 1978), and offers support to the findings of minimal neuron density changes. Unfortunately layers other than the first are not clearly demarcated in the rodent cortex and hence not susceptible to direct measurement.

Diamond et al. (1964) assumed that the total number of neurons was fixed, and that reduction in density was therefore indicative of dilution and could be used as an index of cortical volume expansion. Since the EC occipital depth was greater by 6% and neuron density less by 17%, it was felt that this closely approximated the 19% decrease that might be anticipated if dimensional changes were isotropic. Unfortunately this correlation has not been confirmed, since in the somatosensory region, while depth increased by 4%, neuronal density diminished by only 7%. A more recent study (Diamond et al., 1966), while confirming the EC-IC depth difference of 6%, could detect only a 3% effect on neuron density.

Since the maximum effect occurs in layers II and III, it is noteworthy that EC-IC differences in nuclear and perikaryon diameters were also maximal in these same layers (Diamond, 1967). In the light of these findings it is perhaps significant that layer II increases markedly in width and cell density as the phylogenetic scale is ascended. The results may support Altman's (1967) hypothesis that the neurons of this layer represent the plastic regulatory components of the CNS. The significance of, and the reasons for, the absence of response in layer I and the apparent decrease in thickness of parts of layers IV and V of the EC cortex are unknown. Presumably causes are to be sought in a knowledge of the differential composition and afferent supplies of these layers.

*Neuroglia*

Diamond et al. (1964) found the density of neuroglia per microscopic field to be less in EC than in IC animals, by 6.7% and 5.9% in the visual and somatosensory regions respectively (Diamond et al., 1964). However, since the glia: neuron ratio was increased in the EC animals, and the total number of neurons assumed to be fixed, this finding was thought to indicate an increase in the total number of EC neuroglia. In the individual cortical layers the differences in glial density were so small that no trend could be observed.

In an effort to determine the glial subtypes responsible for the increase, cells were classified by nuclear morphology as astroglia, oligodendroglia, and intermediates (cells which exhibited characteristics of both astroglia and oligodendroglia) (Diamond et al., 1966). The density of glia per microscopic field was 14% greater in the EC than in the IC occipital cortex, and it was found that the proliferation of oligodendroglia and intermediates was responsible for this difference. These cell types showed density increases in the EC brains of 20% and 36% respectively. Examination of astrocytes gave conflicting results: in the first experiment of this paper the EC rats had significantly more (27%, $p<0.01$), but in a replication a slight reversal occurred (-7%, NS). Combining the results of both experiments gave a net increase of 4.8% in the density of astroglia, but this failed to reach significance.

Eleven years later Szeligo and Leblond (1977) confirmed these trends by finding that 30 to 80 days of enrichment increased oligodendroglial numbers by 27% and 33% respectively. The major effect seemed to have occurred by 30 days, but for astrocytes an increase was detectable only in the 80 day subjects, and amounted to 13%. Isolated and social control (3 animals per standard cage) animals did not differ significantly, and in no case were effects noted in the corpus callosum. Handling neonates for 15 minutes daily for the first 10 days of life effected a 12% increase in astrocytes, but not in oligodendroglia. This pattern of glial response differs from the EC-IC one and suggests either an age or a stimulus-specific response; it is interesting in light of the fact that astroglia have been found to proliferate earlier in life, during the preweaning phase, than oligodendroglia, which tend to have a more marked postweaning proliferation.

A similar study by Cummins and Walsh (1978) was only partially successful. Like Diamond et al. (1964), they found that the glia:neuron ration was increased by 8% after 90 days of differential rearing. This was entirely due to a trend towards an increase in astroglia such that the astroglia neuron ratio was increased by 9%, though this increase was nonsignificant. The density of oligodendroglia was 3.6% less in EC than in IC cortex, though once again this trend was not significant, and the number of intermediates counted was too small to allow meaningful interpretation. Likewise no changes were observed in microglia, which are thought to represent blood-derived phagocytic cells and hence to be distinct from other glia (macroglia). For this study, then, it can only be said that despite a suggestive trend towards an increase in astroglia, no significant differences in either macroglia or microglia were detected, and this in spite of very large glial effects in the underlying hippocampus (Walsh et al., 1969). The only other report on cortical histology following complexity-deprivation is that of Altman and Das (1964), who reported an increase in the number of $H^3$ thymidine-labeled cells in the corpus callosum and cortical radiation adjacent to the callosum of EC brains. These cells were initially assumed to be glia, but Altman (1970) now feels that they represent postnatally derived undifferentiated precursors of both neurons and glia.

A complex environment appears to increase cortical glial numbers relative to both social and isolated animals. This increase is greatest for oligodendroglia, which appear to exhibit maximal effects by about 30 days, while astroglia display a greater 80 than 30-day effect. Astroglia also appear to be modified by neonatal handling. Data conflicts on whether or not these effects may also be detectable in the corpus callosum. Although it has been implicitly assumed that the more numerous EC glia represent the result of proliferation, the recent finding by Diamond (1976) that rat cortical glia numbers decrease with aging, between 40 and 108 days of age, suggests that the EC effect might possibly represent a sparing or retention effect.

## Hippocampal Differential Cell Counts

To date the only report of differential cell counts in the hippocampus is that of Walsh et al. (1969), who examined the medial hippocampus underlying the occipital cortex of rats differentially reared for 90 days. The medial hippocampus is composed of two embryologically distinct components, the hippocampus proper and the dentate gyrus. Within both structures neuronal nuclei are aligned into compact and clearly demarcated layers which alternate with layers of neuropil, and within these neuropil layers the counts were performed.

The number of neuronal nuclei in the neuropil is very small, and as might be expected no significant EC-IC differences were detected (Table 2).

For neuroglia, however, the effects were highly significant for all glial subtypes except the microglia. For astrocytes the effect amounted to a 9.8% decrease in density in the complexity animals as compared with isolates.

Since in this experiment the EC hippocampus was of greater size than the EC, the lesser density of astroglia in the EC hippocampus was originally interpreted as a dilution effect due to an increase of internuclear material, rather than to a reduction in absolute numbers. Such an increase in the internuclear material is suggested by the similar trend in occipital cortex (Diamond et al., 1964, 1966) and by the lower density of hippocampal blood vessels which will be discussed shortly. However, further analysis revealed that by far the major component of this difference lies in the Stratum multiformis layer of the dentate gyrus, where other glial and neuronal changes are most evident (Cummins and Walsh, 1978). It may well be that astroglial numbers are indeed altered within this part of the dentate gyrus.

Table 2: Hippocampal Differential Cell Count

Highly significant treatment effects were evident for astroglia, oligodendroglia, and intermediates, but not for neurons or total macroglia. Based on Walsh et al., 1969.

| CELL TYPE | MEAN NUMBER PER ANIMAL EC | IC | PERCENT DIFFERENCE | $X^2$ DF=1 | P |
|---|---|---|---|---|---|
| Neuron | 9.52 | 9.46 | 0.6 | 0 | N.S. |
| Astroglia | 116.6 | 129.4 | -9.8 | 5.658 | 0.02 |
| Oligodendroglia | 29.5 | 21.26 | 38.7 | 11.223 | 0.001 |
| Intermediates | 7.74 | 2.14 | 261.6 | 26.019 | 0.0001 |
| Total Macroglia | 153.4 | 152.8 | 0.4 | 0.001 | N.S. |

The EC-IC differences in the number of oligodendroglia and "intermediates" per microscopic field reached 38.7% and 261.6% respectively (Table 2). In spite of the low number of "intermediates," closer examination revealed that the EC-induced changes in these two glial types was localized to positions which lie within the dentate gyrus rather than the hippocampus proper. Furthermore, there would appear to be a degree of localization even within the dentate gyrus, sicne EC-IC differences were maximal for both cell types within the so-called Stratum multiforme. These findings are interesting in light of Altman and Das' (1964) claim that a proportion of the neurons of the dentate gyrus are formed postnatally from the subependymal layer, and then migrate to the gyrus. If, as these authors have hypothesized, the microneurones so formed represent plastic regulatory CNS components, then the glial changes noted in this study may be associated especially with them. Since oligodendroglia are responsible for myelin formation (Maturana, 1960; Peters, 1960), it would seem possible that heightened axon development occurs in EC animals. It is impossible to determine without the use of serial section, whether these proliferating glia are perineuronal or interfascicular. That glia in the satellite position can proliferate following increased neuronal function, specifically motor activity, has been reported by Kuhlenkampf (1952). An attempt by Diamond et al. (1966) to answer this question by the use of serial sections of the cortex was unsuccessful.

The function of the glial cells with characteristics of both astroglia and oligodendroglia, and even their true identity (whether they are true intermediates in the process of transition from one glial type to another), is unknown. The question is a difficult one, since the very existence of true intermediates is still debated, although work by Moliner (1958), and electron microscopic and autoradiographic evidence (Smart and Leblond, 1961) supports the concept of transitional glia. Brizee and Jacobs (1959) found that spongioblasts continue to divide into adult life and to serve as a reservoir for oligodendrocytes; so the EC conditions may have stimulated precursors to enter the early stages of development and to differentiate into oligodendroglia.

The mechanisms by which an increase in neural activity leads to glial proliferation are unknown. One possibility involves the fact that potassium efflux occurs from discharging neurons, and that glial cell membrane potential and cellular metabolism are highly sensitive to extracellular potassium levels (Orkand et al., 1966). We might hypothesize that the sequence of events and mechanisms mediating glial proliferation, following environmental sensory stimulation in general and complexity in particular, may be one of heightened neural activity modulating the membrane potential of glia and undifferentiated precursor cells e.g. spongioblasts. These changes would in turn elicit alterations in metabolism via genetic derepression and possibly other mechanisms, such as activation of RNA polymerase. The derepression would also induce mitosis. One might predict from this that glial RNA, protein, and size would be augmented following sensory input. Such findings have been reported for glial RNA following experiments involving learning (Hyden, 1967), and might also be predicted for complexity. The functions which might be subserved by an increase in glia are numerous, and include support for newly formed fibers, nutrition, regulation of electolytes, and postulated functions in memory formation.

The above discussion has assumed that the "intermediates" do represent true glia. Altman (1970) has suggested an alternative view: that they may be new or only partially differentiated neurons, at least in the dentate gyrus. It is here that Altman and his co-workers have identified some of the strongest signs of postnatal neuron formation. Such an interpretation would be pleasing, from the point of view that heightened neural activity would be eliciting the formation of additional neurons to process the extra input. From the information available it seems impossible to determine which of these interpretations is correct, but one approach to the question might employ the injection at various intervals of radioactively labeled thymidine, before the sacrifice of differentially reared animals. Newly formed cells would be radioactively tagged and after various periods their migration could be followed. If intermediates moved to the nuclear layers, particularly to the lateral aspect of the Stratum granularis as postnatally derived neurons reportedly do, their neural identity could then be assumed; but if they remained in the neuropil they would more likely be neuroglia. In view of the role of the hippocampus in long-term memory formation, and the implications of new neuron formation in the adult brain, especially in response to the modification of the environment, this would seem an extremely worthwhile experiment.

## Neuronal Morphology

### *Perikaryon And Nuclear Size*

Neuronal size reflects the level of cellular metabolism and as such reflects the intensity, duration and frequency of stimulation (Geinismann, 1971). It therefore seems reasonable to expect variations in sensory input such as complexity deprivation to influence neural size, particularly in view of the evidence for altered metabolism. Diamond (1967) measured neuronal perikaryon and nuclear size in the occipital cortex and found significant differential rearing effects, both preweaning (Malkasian and Diamond, 1971), and 80 days postweaning. The effect was greatest in the upper one third of the 80 day cortex, where the mean EC-IC difference in perikaryon diameter was 16.7% and that of the nuclei was 16.2%. This finding correlates well with neuron count data which indicate that the greatest cortical expansion occurs in the upper layers. Effects were less marked in the lower one third (11.6% and 10.0%) and least in the middle one third (7.2% and 7.6%). Thus the main increase in both amount of internuclear material (Diamond et.al., 1964) and cell size was in the outer one third of the cortex. There was also a positive correlation between nuclear size and cortical depth. However, when the duration of differential rearing was reduced to only 30 days, it proved insufficient to elicit significant effects on nuclear size. This was true whether the EC-IC program was begun either at weaning or at 60 days of age. For this reason measurements of perikaryon dimensions were obtained, and resulted in a detectable difference (4.5%), which confirms previous findings that nuclear size may not always correlate with that of the perikaryon (Zetterberg, 1966).

Another form of environmental stimulation which might be labeled as complexity was used by Schapiro and Vukovich (1970). They subjected neonatal rats to a thrice daily 20 minute combination of tactile, thermal, auditory and visual stimuli from day one to eight, and examined pyramidal neurons in the occipital cortex after rapid Golgi staining. No effect on perikaryonal size could be detected even though dendritic spine numbers were increased. However, only a small number of neurons in five pairs of animals were examined, which probably would be insufficient to detect differences of the magnitude reported by Diamond (1967). The reliability of the rapid Golgi technique for cell size measurement is also unknown, so that it is difficult to interpret this report.

The only other information on neuronal size comes from Walsh and Cummins (1979). They attempted to follow up their finding of hippocampal glial changes by measuring neuron nuclear size in the hippocampus and dentate gyrus after 80 days of rearing. Both maximum and minimum nuclear diameters were measured and the ratio of the two computed, as well as estimates of nuclear area and volume. In addition to the mean value in each animal for each of these parameters, their variability within individual animals was determined by the coefficient of variation (standard deviation divided by the mean). Interestingly enough, while no overall differences in size could be detected, environmental effects on variability were evident for all parameters. The effect was such that the coefficient of variation was significantly greater in IC than EC animals. However, this effect was evident only in the Stratum granularis of the dentate gyrus. This is the same layer to which postnatally formed neurons have been reported to migrate (Altman, 1970), and lies adjacent to the Stratum multiformis where glial changes are most marked. The effect on the ratio of diameters suggests more than a simple overall change in size; an actual effect on nuclear shape resulting in increased pleomorphism (variablility in shape). One interpretation might be that the greater variation in isolates reflects a more heterogenous cell population-possibly due to a greater formation of new neurons. But this runs counter to the previous suggestion that the greater number of intermediate glia in the adjacent Stratum multiformis of the EC hippocampus reflects an increased number of undifferentiated or nascent neurons migrating to the Stratum granularis following greater proliferation under complexity. Perhaps newly formed neurons represent a more homogenous, less differentiated picture than older ones. In this case the results would reflect greater numbers of new neurons in the EC brains.

Another explanation of a more heterogenous IC neuron population in this layer might be that the decreased stimulation and hence neuronal activity are such that differentiation and maturation of newly arrived cells are delayed. If differential effects on new cell formation and migration are not a factor, then perhaps that lack of stimulation stunts maturation of some neurons in this layer, such that they do not approach their genetically determined ceiling levels. This might occur as a primary neuronal response to low levels of stimulation, or might reflect a greater variability of input to individual cells, due perhaps to the formation of fewer well-functioning pathways. Neurons involved in these pathways are highly stimulated, while those not so innervated are hypoplastic. The difficulty with these last explanations is that they do not account for the effect on the Stratum granularis with no effect on the other layers, unless neural activity follows a similar regional pattern. This is suggested by the proximity of the glial changes, but here again we are up against the problem of our current ignorance concerning the precise patterns of neural activity which follow rearing under complex and deprived environments. Without this knowledge it is extremely difficult for us to be certain of the interpretation of any regional and temporal patterning of anatomical and biochemical changes. Such factors may even be particularly important for neuron size, since it has already been shown that the duration, intensity and character of stimulation determine the direction, extent and general pattern of volume changes (Geinismann, 1971). In any event, this study is interesting in that it raises the posibility that the environment may be capable of modifying the process of postnatal neuronogenesis. To date there have been suprisingly few studies of neural variability, and the only other report on coefficients of variation comes from Diamond et al. (1975), who did not observe any significant effects on cortical neuron nuclear size.

In any event, in view of the large amount of work involved in simply performing the measurements of neuron size, it would seem a worthwhile investment to add the relatively simple calculations to determine variability as well as mean size in future experiments.

## Dendritic Morphology

"One might suppose that cerebral exercise, since it cannot produce new cells, carries further than usual, protoplasmic expansion and neural collaterals, forcing the establishment of new and more extended intercortical conections" (Cajal, 1895).

Over 70 years were to pass before the experimental validation of Cajal's hypothesis. In their original report of increased cortical thickness, Diamond et al. (1964) suggested increased dendritic ramification as one of the factors contributing to the greater EC cortical bulk. They based this hypothesis on a report that postnatal cortical development is due largely to dendritic proliferation (Eayrs and Goodhead, 1959), and that there is a correlation between neuronal perikaryonal size and dentritic branching (Clendinnen and Eayrs, 1961).

## *The Dendritic Tree*

The first report on the size of the dendritic tree was Holloway's (1966), who measured with a series of concentric circles placed around a neuron, the number of intersections that dendritic branches made. He reported that after 80 days the layer II EC occipital stellate neurons made more intersections with the peripheral circles than their IC counterparts.

Unfortunately Holloway's method does not tell whether this difference is due to variation in length or branching or both. However, Volkmar and Greenough (1972) suggest that branching is definitely involved. They reared Long Evans hooded rats from weaning for 30 days under conditions of EC, IC, and pairs in IC cages. In the medial occipital cortex the number and order of dendritic branching of layer II, IV and V pyramidal cells were measured. In EC all cell types displayed a greater branching of higher order, but not basal, dendrites, with the effect being most marked in the more peripheral branches. Compared with isolates, paired subjects showed only small effects in the most distal branches, whereas EC effects were quantitively larger and detectable in intermediate as well as distal branches.

Greenough et al. (1973) next asked whether these findings were unique to the occipital cortex or could also be detected elsewhere. To answer this question, pyramidal cells in layers II and IV of frontal cortex and layer IV and V of temporal cortex were examined and three different types of parameters obtained. First, the intersections between dendrites and each of a series of concentric circles at 20μ intervals from the cell body were counted (Sholl, 1956). This measure is sensitive to the combination of number, length, and location (with respect to the cell body) of dendritic branches but less sensitive to the differences restricted to a small portion of the neuron. Second, the number of branches at each order away from the cell body (or in the case of pyramidal cell oblique dendrites, from the apical shaft) were recorded. A branch from the cell body or apical shaft was defined as order 1, both branches past the first bifurcation as order 2, and so forth. This analysis is more sensitive to the differences in higher order dendrites which the authors previously found in occipital cortex following differential rearing. Third, the projected lengths of dendrites of each order were recorded for the frontal cortex neurons.

No systematic pattern of environmental effects on dendrites could be detected in the frontal cortex. On the other hand there was a clear-cut effect of the litter from which the subject had come. Similarly for the temporal cortex, analyses of apical branching by order, then of whole neurons by the intersection of dendrites with rings, showed only occasional effects. However, environmental effects were clearly apparent in the higher order branching of basal dendrites. In layer IV the complexity animals had significantly more basal branching than their isolated and socially housed littermates, while in layer V both environmental complexity and isolated animals had more basal branches than social rats. As in the frontal cortex there were significant differences among litters on all measures.

The effects in this study, though much smaller than those found in occipital cortex, are qualitatively similar in that the environmental effects appear to be restricted to basal dendrites, as are effects on spine density (Globus et al., 1973). Since these effects are not seen in all neuronal populations, it seems doubtful that the environmental effects on dendritic branching are mediated by systemic factors such as hormonal or nutritional differerences. The specific factors involved await further elucidation.

This still leaves unanswered the question of whether length as well as branching is affected, but the work of Globus et al. (1973) suggests that environmental effects on dendritic length, if they exist at all, are quite small. They measured the lateral width of occipital cortex pyramidal cells in S1 strain rats which had been differentially reared for 30 days from weaning. The lateral width and length of a pyramidal cell are measures of the diameter and length of the cylinder whose surface is defined by the ends of the terminal, oblique and basal dendrites (Globus et al., 1973). EC and IC width values differed by 1.4% but this was not significant.

## *Dendritic Spines*

Globus et al. (1973) also performed counts of the number of spines per unit length of dendrite and found significant effects. Complexity produced an increase in spine numbers which was most marked in the basal dendrites (9.7%), smaller though still significant in the oblique and terminal branches (3.6% and 3.1% respectively), and apparently absent in the midsection portion of the apical dendrite. This finding is similar to that of Schapiro and Vukovich (1970) who, as previously mentioned, subjected neonate rats to thrice daily multimodality stimulation for eight days from birth. Under these conditions all types of dendritic branches displayed increased spines with maximum effect (33%) in the oblique and basilar, an intermediate amount (28%) in the terminal, and least (10%) in the apical. These effects are considerably larger than those of Globus et al. (1973), a fact which the latter attributed to the younger age of Schapiro and Vukovich's subjects and the stressful nature, e.g., mechanical vibration and electrical shock, of some of the stimuli used. Independent support for these findings is available from studies of light deprivation and cortical electrical stimulation, both of which have been found to elicit a variety of dendritic responses (Valverde, 1971; Rutledge, 1976).

The patterning of effects among the various branches is interesting in view of their principal connections. Thus apical branches are said to receive most afferents from the specific afferent radiation, which in the case of the visual cortex is from the lateral geniculate; it is these spines which are most affected by dark rearing (Globus and Scheibel, 1966, 1967a; Valverde, 1968). These specific afferents would probably be little affected by environmental complexity or deprivation, since the total amount of light input is probably about the same in both conditions. Most likely it is the cortical integrative functions which are most called into play by EC as opposed to IC conditions. Thus it is interesting that the apical branches show least effect and the basal, which receive primarily intracortical connections, display most. The oblique branches, which were the most affected in the Schapiro and Vukovich (1970) study but very little affected in the Globus et al. (1973), receive most of their afferents from the interhemispheric corpus callosum fibres (Globus and Scheibel, 1967b). Perhaps interhemispheric connectivity is more susceptible to modification in the neonate animal.

Studies of dendritic spines, employing light deprivation and stimulation, suggest that there may be three classes of spines which can be differentiated according to their responses to visual experience. The first develops independently of stimulation: the second is dependent on stimulation though this stimulation may occur at any time: and the third requires stimulation at a specific developmental period. This finding comes from Valverde (1971); Valverde and Ruiz-Marcos, (1969), who examined spines on the apical dendrites of pyramidal cells in the visual cortex of mice. The stimulation independent spines developed around the time of eye opening at about 12 to 15 days of age, whereas the second class appeared later whenever the light deprived animals were exposed to light. The third group of spines developed only when the animals were light exposed at the time of eye opening, that is, both age and experience dependent. The "critical period" class of spines correlates well with the behavioral finding that the normally light exposed animal may reach a level of perceptual sensitivity which the deprived animal cannot achieve. The second class of neurons (stimulation dependent) is particularly interesting from a perspective of memory formation, since it suggests the existence of a population of neurons that appears to be waiting to be modified or be made functional by experience. Similar neurons could conceivably be waiting to encode experience in other brain regions (Greenough, 1976).

*Effects of Formal Training On Dendrites*

Following their discovery of the effects of social and enriched rearing on cortical dendrites when compared with animals reared in isolation, Greenough (1976) attempted to determine whether formal training could elicit similar effects. They therefore trained rats in two different types of procedures: the first a two-alternative visual discrimination task, and the second a Hebb-Williams maze. Neither procedure brought about the large increases in basilar dendritic branching which they had formerly detected in the environmental studies. Rather, they noted that oblique and/or terminal branches from outer regions of layers IV and V pyramidal cell apical shafts were longer in animals who had received the extensive formal training. Compared to the environmental studies, the effects were more subtle and located at the outer extremities of the dendritic tree.

*Mechanisms Mediating Changes In Dendritic Spines*

Since it is now clear that environmental stimulation is capable of modifying both spine numbers and morphology, the obvious question arises as to the mechanisms involved. A possible clue is available from the work of Van Harreveld and Fifkova (1975), who noted swelling in spines immediately following stimulation. They examined hippocampal granular cells and compared spines on the same dendrites where some spines were known to have been stimulated and others not. Thirty seconds of stimulation resulted in an enlargement of the cross-sectional area of the stimulated spines, which was present two minutes after stimulation and had not noticeably decreased after 60. Swelling resulted in a decrease in the spines' electrical length resistance and thus seems one possible mechanism to account for post-tetanic potentiation. If this swelling is stabilized by more enduring processes, such as the laying down of structural proteins, then it might also provide a basis for long-term morphological and functional effects. Van Harreveld and Fifkova suggest that this post-tetanic swelling may be similar to the swelling which has been observed following spreading depression (Van Harreveld and Khattab, 1967). The uptake of electrolytes and of water which causes the swelling seen during spreading depression, is thought to be due to a release of glutamate from the intracellular compartment into the extracellular space, which suggests that a similar mechanism may possibly underlie post-tetanic potentiation and environmentally induced effects.

*Summary*

Thus it is evident that Cajal's (1895) predictions have been borne out. A complex environment increases both dendritic arborization and numbers of spines in both stellate and pyramidal cells, but the effect on the length of individual dendrite branches is still uncertain. There is a specific patterning of such changes: branching occurs in the more peripheral portions of the tree, and the changes in spine density are greatest in those branches subserving intracortical connections and least in those receiving specific sensory afferents. Whether, like spine growth, environmentally elicited bifurcation occurs only in specific types of dendritic branches, is not known. Presumably the greater arborization, plus the increased density of spines, reflect axonal arborization and greater synaptic numbers.

Synapses

After decades of speculation (Cajal, 1895; Hebb, 1949) on the possible effects of environmental enrichment on synaptic morphology, the first experimental support was provided by Mollgaard et al. (1971). Following 30 days of EC-IC rearing, layer III neuropil of the occipital cortex was subjected to electromicroscopy, and the size and number of Gray's (1959) type 1 synapses were measured. Type 1 synapses are associated with dendritic spines and identified microscopically by the presence of round vesicles in the presynaptic bouton and asymmetrical cytoplasmic thickening of the postsynaptic membrane. Hence they are sometimes referred to as asymmetrical axodendritic synapses.

EC postsynaptic length was 52% greater but synapses were only 67% as numerous per unit volume of cortex. That the increase in length was not due simply to a loss of small synapses was demonstrated by the fact that the EC animals possessed more large synapses, as well as fewer small ones, than did their littermates. It was estimated that the net effect of these changes was to increase total synaptic area in the EC group by the order of 40%. In this experiment cortical depth was found to be increased by 4%, and the EC-IC littermate differences in depth correlated closely with synaptic length differences ($r=0.79$), but not with synaptic numbers. Replication studies in the Berkeley labs have resulted in similar though somewhat smaller effects, with SC values lying close to complexity ones (Diamond et al., 1975b).

These findings have been extended to other occipital layers by West and Greenough (1972), who measured bouton diameter and length of the postsynaptic thickening of type I synapses. After 30 days of differential rearing, the postsynaptic thickening was significantly longer in EC than IC cortical layers I and IV, but not VI, while the diameter of the presynaptic bouton was not appreciably altered. Because synaptic numbers were unfortunately not counted, it is not known whether the inverse relationship between size and number holds in this case.

It should be noted that the findings of reduced synaptic numbers per unit volume in the EC cortex stand in contrast to the report of increased dendritic fields and dendritic spines, which findings would lead to expectations of greater numbers of synapses. The findings may reflect a dilution effect within an expanded EC cortex. In addition, Cummins (1978), working with mice, has recently replicated his findings of greater synaptic numbers in the EC cortex (Cummins et al., 1975; Walsh and Cummins, 1976c) and, in all-too-rare example of interlaboratory cross-validation, an examination of his photographs by Marian Diamond suggests that they are employing equivalent criteria for synaptic identification.

Postsynaptic thickenings sometimes display gaps or perforations, called subsynaptic plate perforations, which are more frequent in larger synapses. Greenough et al. (1978) wondered, therefore, if the larger synapses of EC and SC animals might not display more perforations than the smaller IC synapses. This was in fact the case and both EC and SC rats had an average of 25% more synapses in all cortical layers than did the IC rats. The functional significance of this shift is uncertain but in light of the greater frequency of perforations in the more experienced group, the authors suggest that they might be associated with positive functional changes. SC values appear to be close to those of EC.

Clearly the EC-IC conditions modify synapses, but the final story on exactly how remains to be told. The net effect is probably to increase the total synaptic interface between neurons and presumably therefore to augment connectivity. Whether other components of the synapse also enlarge to support this increase in activity, and whether other types of synapses are similarly affected, remain to be seen.

## Mechanism Mediating The Modification of Neuronal Interconnections

It seems clear that experience is capable of modifying the numbers and morphology of dendrites and synapses, and probably also of axons. The mechanisms which mediate these effects are by no means so clear. Alterations in the amount of axonal and dendritic growth by modification of the appropriate metabolic processes are not difficult to imagine, but the question of how this growth would be directed towards appropriate neurons is not so obvious. Various chemical affinity mechanisms have been suggested for neural development, and are probably essential for inducing basic developmental patterns of brain organization; but whether such a specification could include new information gained from experience is less certain. Perhaps electrically active dendrites might secrete a chemotropic substance which would attract growing axons. Moore (1976) has suggested that spike activity might cause axons to sprout extra terminals, which would then compete for nearby synaptic sites.

Synaptic modifications have been suggested to be of several types, including new synapse formation, changes in size and morphology, loss and preservation of pre-existing synapses. The preservation hypothesis has been gaining support recently from a number of areas, and basically contends that the brain and individual neurons initially make many temporary or tentative synaptic contacts, and that only a proportion of these actually survive. Survival is supposed to depend on their functional utility, which in some cases would be partially determined by experience. Presumably fewer synapses would be preserved in stimulus-deprived than in enriched animals. Several authors have suggested similar but not completely identical models (Rosenzweig et al., 1972; Changeux and Danchin, 1977; Greenough, 1978). The following discussion owes much to Greenough (1978).

Support for this preservation hypothesis comes from time course studies of both neurons and synapses. For example, in the frog spinal cord it appears that about four times as many neurons involved in limb muscle innervation are produced as survive to adulthood. This survival appears to depend upon a neuron's successfully competing with others to innervate a muscle (Prestige, 1970), and Hughes (1968) has suggested that this process may allow the selection of the most appropriate neurons for an organized neuromuscular control system. Indeed, selective survival may be commonplace in neural development, since neuron death occurs in a variety of places in the developing central nervous system, including the cortex (Diamond, 1976).

A similar pattern may occur for synapses. For example, Ruiz-Marcos and Valverde (1969) report that the number of synapses in mouse cortex reaches a peak shortly after weaning and then begins to fall toward an adult level, and that this final level is lower still in dark reared mice. The same picture occurs in the cat's visual cortex where synaptic density peaks near the end of the fifth postnatal week (Cragg, 1975). This peak matched the period of maximal sensitivity to monocular deprivation (Hubel and Weisel, 1970). Westrum (1975) described multiple axon contacts on single dendritic sites in early development, but only one axon terminal appeared to survive to maturity. Competition for synaptic sites has also been proposed in other developing systems, especially the visual where it is the basis of many of the monocular deprivation effects (Hubel and Wiesel, 1970).

It is probably simplistic to think of neuronal interconnections in the adult brain as fixed and static. Probably a more adequate picture is provided by a dynamic model in which axonal, dendritic, and synaptic growth is in constant dynamic flux. The net result would approximate homeostasis, unless sensory input or other modifying factors resulted in a shift toward another homeostatic balance point.

Blood Vessels

Changes in blood flow and vascularity have been found following a variety of forms of sensory stimulation and neural activity. It seems reasonable to expect similar effects following exposure to complexity and isolation. To date, however, only two reports, both of blood vessel counts, have appeared, and the effects are not yet completely clear. Both Diamond et al. (1964) and Walsh et al. (1969) differentiated vessels on the basis of apparent size, into those of capillary dimensions (less than 5 μ ), and those larger. After 80 days of differential rearing Diamond et al. found that the occipital cortex density per microscopic field in the EC, as compared with the IC, was less by 12% and greater by 26% for small and large vessels respectively. A similar though smaller and nonsignificant trend was evident in the somatosensory cortex. Walsh et al. performed a similar study after 90 days but reported a trend toward a lower EC density for both small (-5.7%) and large (-15.6%) vessels in all layers except II and IV, thus supporting Diamond et al. as regards capillaries, but not larger vessels. This conflict might possibly reflect different histological procedures, since the Berkeley workers used 25 μ sections as opposed to 7μ sections in the Australian lab. The thicker sections would allow full visualization of the complete width of a greater percentage of large vessels, and hence greater sensitivity in determining their numbers. Walsh et al. also made counts in the Strata molecularis and radiatum of the medial hippocampus, finding here also both large (-8%) and small (-8.5%) vessels to be less dense in the EC brain. Taken separately, the effects for occipital cortex and hippocampus were not significant, but when combined were significant at the 0.02 level, for small but not larger vessels (Cummins and Walsh, 1978). This reduced density of capillaries in the EC animals could possibly be attributed to dilution in the expanding complexity brain.

It seems that the density of capillary-sized vessels is less in EC than IC cortex, and perhaps hippocampus, while the fate of larger vessels is uncertain. Thus the apriori assumption of greater blood flow and vascularity in the EC brain, to accommodate an increased metabolism, has not yet been confirmed.

## PHYSIOLOGY

### Factors Contributing To The Differential Development Of Body Weight

The effects of differential rearing are not confined to the brain but affect a wide range of somatic systems. One of the most commonly noted findings has been a differential development of body weight, with isolates being heaviest, complexity subjects lightest, and social animals intermediate. An exception to this general principle is that in longer term studies, exceeding 100 days, the social animals tend to become heaviest (Rosenzweig et al., 1971).

An obvious question concerns the factors which contribute to this differential development. Some of the more obvious contenders include the skeletal, muscular, and adipose systems, as well as certain specific organs. Although to date there appear to have been no investigations of muscle mass, the other contenders have been examined. While Fiala et al. (1977) reported that informal observations failed to detect differences in skeletal size, a more precise quantitative analysis in the Berkeley labs (Diamond et al., 1965) noted significant differences in the dimensions of the skull. It therefore seems quite possible that differences in the remainder of the skeleton might also be detectable by precise measurement.

A series of experiments involving over 400 subjects revealed consistent differences for a variety of internal organs (Cummins and Walsh, 1978). Isolates at a variety of ages consistently had heavier liver, spleen, adrenal, testicular, heart and pituitary weights.

Further informal observations by Fiala et al. suggested that isolates have considerably larger deposits of adipose tissue. They therefore conducted a study of food and water intake to determine whether this could account for the findings. In both male and female rats differentially reared from weaning for four weeks, isolates consistently consumed greater amounts of food and male isolates consumed significantly more water. Similar observations of food intake have been noted by Tagney (1973). Multiple factors may account for this differential intake, including excess eating by the isolates because of boredom and the lack of anything else for them to gnaw. It has also been noted that animals must obtain at least some minimal level of exercise before they begin to regulate food intake in accordance with need. Probably some combination of these factors and mechanisms will provide the most comprehensive explanation.

### Electrophysiology

#### *Cortical-Evoked Potentials*

Cortical-evoked potentials are thought to represent effects of both direct stimulation and cortical processing of sensory stimuli, and as such might be expected to display effects of differential rearing. Edwards et al. (1969) found that after 8 weeks differential treatment, the EC mean latency was less than IC (56 msec vs 64 msec). Mailloux et al. (1974) then expanded this study to try employing three environments: visual enrichment, auditory enrichment, and a social control group, on subjects differentially reared from weaning to 95 days of age. They then recorded photic- and auditory-evoked potentials in unanesthetized animals from chronically implanted epidural electrodes over the visual and auditory cortex, and the latencies of these potentials were measured. Only those animals whose enrichment included visual stimulation displayed significant effects of reduced latency, and even here the effect was limited to visual cortex latencies recorded following photic-evoked responses. No significant effects could be determined in auditorially enriched animals or in the auditory cortex of animals exposed to either visual or auditory enrichment.

In some ways this finding is anomalous, since morphological and biochemical effects of enrichment have been detected in nonvisual cortex even in dark reared, blinded, or anosmic animals. This suggests that there may have been other effects that failed to reach significant levels. On the other hand, it may be that evoked potentials are more indicative of immediate sensory modality specific effects than of less immediate and less modality specific general cortical activation. Then again, if direct interaction with stimulus objects, and consequent reafferent stimulation, are the important factors (Rosenzweig and Bennett, 1976; Walsh and Cummins, 1976a), the auditory stimuli might then be expected to be relatively ineffective.

## Sleep

Since increases in rapid eye movement (R.E.M.) sleep have been noted in rats after exposure to learning situations (Lucero, 1970; Holdstock and Franks, 1971), Tagney (1973) surmised that EC rats, presumably having more to learn, might display greater R.E.M. sleep than their isolated counterparts. She therefore performed an electrophysiological analysis on first generation crossbred rats which had been differentially reared for approximately 30 days. First generation crosses were used, in light of Henderson's (1970) finding that brain weight effects were greater in these than in inbred mouse strains. EC rats displayed significantly more total sleep time than isolates, a difference seen in both slow wave and R.E.M. sleep times. Thus the percentage of R.E.M. sleep was the same in both groups.

After 80 days in their respective environments, five IC rats were transferred to the complex environment, and after a further 16 days were recorded. These transferred animals displayed significantly increased slow wave sleep while R.E.M. displayed a similar, though smaller and not significant, trend.

Tagney interpreted these findings in the light of several kinds of evidence which suggest that sleep is the time in which the brain may be especially effective in synthesizing macromolecules (Feinberg et al., 1969; Oswald, 1969; Parker et al., 1969). She dismissed the possibility that differential amounts of physical activity could account for the finding, since Webb and Friedman (1971) could detect no overall alteration in the onset of sleep in rats having chronic access to activity wheels. She therefore concluded that the increased sleep occuring in EC rats was due to their increased cerebral activity, and consequent requirements for longer periods of optimal conditions, for restorative macroprotein synthesis.

Other studies report similar trends. Isolation-reared kittens slept less than group-reared and handled ones (McGinty, 1971). When the isolates were exposed to complexity, the percentages of time spent in both R.E.M. and slow wave sleep increased. A further study clearly suggested that it was the novel stimulation that was the causative agent.

Implanted electrode recordings from rats differentially reared for six weeks revealed that EC inhibited not only total sleep time but also, contrary to Tagney, R.E.M. (Lambert and Truong-Ngoc, 1976a, 1976b). The EC rats also had a higher threshold of activation of cortical responses following reticular formation stimulation. However, reticular excitability and sleep behavior were not correlated, and it was concluded that the environment had independent effects on these two phenomena.

## NEUROCHEMISTRY

Considerable effort has gone into analysis of the biochemical processes underlying the observed weight changes. Most interest has centered on substances thought to be associated with neurotransmission, particularly the cholinergic system. This reflects largely the early findings (Krech et al., 1960; Rosenzweig et al., 1962; Thiessen et al., 1962; Zolman and Morimoto, 1962) of regional specificity of changes in cholinesterase, together with absent or very small changes in protein and RNA concentration. Recently, more precise studies of RNA and protein composition and activity have begun to emerge with positive findings.

### Energetics

To date there is apparently no published information on the effects of differential rearing on the brain's energy requirements, but preliminary data of Ferchmin and Ferchmin (1973) suggest an increased *in vitro* oxygen consumption by cerebral cortex following EC as would be expected from the greater size and metabolic activity of the EC cerebrum.

### Neurotransmitters And Related Substances

#### *The Cholinesterases*

The original finding of a complexity-associated decrease in ACh hydrolysis per unit weight in the cortex, accompanied by an increase in the subcortex, has since been replicated and considerably extended. Over 95% of ACh hydrolytic activity in the rat brain is due to the specific enzyme acetylcholinesterase (AChE), as opposed to the nonspecific cholinesterase (ChE) (Bennett et al., 1964). These enzymes respond differently, but both display a pattern of regional specificity in some ways analogous to that of weight.

Eighty days of environmental complexity increase the total activity of both enzymes in all areas, with the exception of somesthetic cortex for AChE and subcortex for ChE (Rosenzweig et al., 1971). Within the cortex the pattern is similar to that of weight, increases being most marked in the occipital region and least in somesthetic (Rosenzweig et al., 1969). However, the effects are much more marked for ChE, and the ratio of ChE to AChE activity is greater in EC animals in all regions except the subcortical.

Enzyme activity can also be expressed per unit weight of tissue; it is this method which most clearly differentiates the ChE and AChE responses. For AChE the increase in total activity in cortical regions is less than the corresponding weight increase, so that AChE per unit weight (AChE/wt, concentration) is actually decreased. But in the subcortex, where total amounts are unaffected and weight falls, the reverse occurs and AChE/wt is increased. For ChE, on the other hand, the cortical increases are greater than those of weight. In the subcortex total ChE is unaffected, but weight falls so that cholinesterase concentration, or ChE/wt, is increased in all brain regions.

The regional specificity of changes within the subcortex has been studied in detail only once, and then only total ACh hydrolytic activity was assayed (Rosenzweig et al., 1962). Only the cerebellum displayed significant effects, and elsewhere they appeared uniformly small. While regional specificity probably exists„ the size of subcortical changes is too small to allow its easy detection. As with weight, the opposite directions of the cortical and subcortical enzyme activity/weight changes yield a large and more stable C/S ratio than either of the primary factors alone.

In addition to the general nature of responses, the regional specificity of the weight and enzyme changes have also been confirmed by several workers (e.g., Zolman and Morimoto, 1962; Geller et al., 1965; Reige and Morimoto, 1970). While the magnitude of the effects may vary, the finding of regional specificity would seem to be general across strains and sex, at least as far as the rat is concerned (Krech et al., 1960; Rosenzweig et al., 1969). This is not necessarily true for other rodents, since the gerbil exhibited a more uniform cortical change and a less marked C/S ratio effect (Rosenzweig and Bennett, 1969), while in the mouse the greatest effects were in the ventral and somesthetic areas (LaTorre, 1968). Essman (1971) reported that mice exhibited huge (approximately 50%) changes in both AChE and ChE. This differs markedly from all other studies and will require replication before being accepted. It would be interesting to extend these cross-species comparisons to subjects higher on the phylogenetic scale, especially to primates.

What then is the functional significance of the AChE and ChE changes widely reported to follow environmental complexity and isolation? AChE measurement was designed as an index of synaptic activity (Krech et al., 1962), since it is largely localized to neuronal synapses in the rat brain. In addition to synapses, however, AChE is also found in non-synaptic neuronal cell membrane (DeRobertis, 1971) and endoplasmic reticulum (Torack and Barnett, 1962; Mori and Shimizu, 1964) so that it is perhaps better regarded as an index of synaptic and other neuronal activity. Although its neural function apparently remains unknown, ChE is highly localized to glia (Koelle, 1952, 1954; Giacobini, 1959), and its increase has been used as an indicator of glial proliferation (Bennett et al., 1964). While it is not yet known how precisely these enzymic changes correlate with their histological substrates, and although their precise functions remain unknown, microscopic evidence of the predicted synaptic and glial changes have been identified.

## Other Cholinergic Substances

In view of the greater difficulties involved in assaying them, ACh itself and its synthesizing enzyme, choline acetyltransferase (ChAt), have been little studied. The only available report on ACh (Rosenzweig et al., 1971) failed to show an EC-IC difference but this is hardly surprising since the procedure necessitated whole brain assay and involved a relatively insensitive bioassay. ChAt on the other hand was found to follow a pattern similar to AChE, exhibiting a slight increase in total cortical activity but relatively less than the concomitant weight rise.

## *Biogenic Amines*

The other major group of proposed neurotransmitters, the biogenic amines, have been much less closely studied in environmental complexity than have the cholinesterases. In part this reflects the early success of the Berkeley team in correlating cholinesterase levels and problem solving ability, and their initial lack of success in identifying serotonin changes in the cortex (Bennett et al., 1964).

Similarly, Geller et al. (1965) were unable to detect effects on whole brain serotonin, or dopamine, or the decarboxylase enzymes of 5-hydroxytryptophan and glutamic acid. They did, however, note a reduction in the amount of norepinephrine in the EC brains. These findings have been extended by Riege and Morimoto (1970), who found an EC norepinephrine reduction in dorsal cortex and diencephalon-mesencephalon areas, but an increase in the third region examined, the hypothalamus. In addition, cortical serotonin and hypothalamic dopamine were lowered. Essman (1971a) has differentiated "free" and "bound" serotonin and reports that it is only the former which is lowered by aggregation and stimulation. The regional specificity of these responses explains why whole brain assays such as those of Geller et al. are unlikely to be successful.

In view of the effects of a variety of stimuli such as electric shock (Thierry et al., 1968), isolation (Geller and Yuviler, 1968) and population density (Giacalone et al., 1968) on biogenic amine turnover rates, similar effects will almost certainly be found for environmental complexity. However, their interpretation at the organismic level, e.g., for behavior may be difficult, given the present stage of limited knowledge of the functional significance of those substances.

## *Cyclic AMP*

In recent years there has been considerable interest in the cyclic AMP system, which appears to be a final common metabolic pathway mediating the effects of a variety of hormones on their target cells. Hormones appear to activate the enzyme adenyl cyclase with consequent increased production of cyclic AMP and other adenosine nucleotides which appear to adapt to chronic alterations in synaptic input. Thus when cortical levels of the neurotransmitter norepinephrine are chronically depleted, as by reserpine, 6-hydroxy-dopamine, or lesions of the medial forebrain bundle, then the post synaptic norepinephrine sensitive cyclic AMP systems become hyper-responsive. It therefore occurred to Dismukes and Daly (1976) that measuring the responsiveness of cyclic AMP generating systems in the brain might provide a valuable method for determining the effects of environmental manipulations on the synaptic activity of different neurotransmitter pathways in the central nervous system. They therefore examined the accumulation of radioactive cyclic AMP, following administration of various neurohormones to the brains of EC and IC rats which had been differentially reared for 30 to 60 days. Sections of the telencephalon were incubated with five agents which produced moderate to large stimulation of cyclic AMP formation in rat brain slices. No significant differences were found between EC and IC brains in basal levels of cyclic AMP, or in the accumulations of cyclic AMP elicited by maximally effective concentrations of either norepinephrine, isoproterenol, or adenosine. However, Prostaglandin $E_1$ (P.G.E.$_1$) produced a 60% greater accumulation of cyclic AMP in IC than EC brain slices. These values appeared comparable to those previously observed with standard colony rats. This effect did not seem to represent a differential incorporation of radioactivity into precursor compartments, but rather to be indicative of either differential activity of prostaglandin sensitive adenyl cyclases or decreased catabolism of cyclic AMP by phosphodiesterase. A nonsignificant trend towards greater cyclic AMP stimulation by histamine in EC brains was also noted.

These differences in responsiveness of cyclic AMP generating systems suggest that environments may chronically alter the functional synaptic activation of adenylate cyclases responsive to certain putative neurohormones. This type of picture is not unique, since alteration in environmental lighting has been shown to control noradrenergic input to the pineal gland, producing corresponding alterations in responsiveness of the postsynaptic pineal cyclic AMP system (Axelrod, 1974).

## Nucleic Acids

Despite sporadic claims to the contrary, it seems well established that total DNA is largely unaffected by sensory stimulation or deprivation (Pevzner, 1966) and this has been confirmed for environmental complexity

(Rosenzweig et al., 1971). This is not so for RNA, which is one of the earliest and most sensitive biochemical indices of stimulation. Initially there were few studies of environmental complexity and isolation effects on this substance, since early reports suggested that changes in concentration did not occur, and that total amounts merely reflected weight changes (Bennett et al., 1964; Rosenzweig, 1966). However, Ferchmin et al. (1970), using larger numbers of subjects, have been able to identify changes not only in total cortical RNA but also in concentration. Since total DNA remains relatively unaffected by environment, the concentration in the expanded complexity cortex is reduced. Thus the ratio of RNA concentration, which increases, to DNA concentration, which decreases, yields perhaps the most reliable chemical index of environmental effects (Rosenzweig and Bennett, 1978). The temporal pattern of these changes will be discussed in more detail in a subsequent section.

The only detailed study of subcortical RNA has been Essman's (1971), previously mentioned with regard to somatosensory cholinesterases. In this same study Essman reported that giving toys, handling, and other intermittent stimulation for 15 days, to mice which had already been either isolated or socially reared for 28 days, resulted in marked changes in a "glia enriched" cerebellar fraction. Aggregation was without effect but stimulation increased the weight of the fraction by over 100%. For the RNA concentration of the fraction, stimulation interacted with aggregation since it affected only isolates, with stimulated isolates having higher, and non-stimulated isolates having lower, levels than aggregated subjects. The interpretation of these findings is problematical since other studies of cerebellum have examined only weight and cholinesterase and detected only very small changes (1 - 2%; Rosenzweig et al., 1962). The effects are some 20 times greater than those of Ferchmin et al. (1970) for the cerebrum, but Essman reports directionally opposite changes for glial and neuronal RNA, a factor which would greatly reduce the sensitivity of a total RNA analysis such as Ferchmin's. Essman's study has been reviewed in some detail because it stands out so markedly from all other work in the field, by virtue of the sheer magnitude of the effects reported and the interaction between social and non-social stimuli. It is certainly a study which warrants attempts at replication.

## Gene Activity

Within the differentiated eukaryotic nucleus, DNA exists in a complex in association with protein and RNA. This DNA complex when extracted from the nucleus is known as chromatin. Although all differentiated cells contain a complete copy of the genome, only a limited portion of the DNA is actually transcribed. It is widely agreed that the protein moieties associated with DNA are important regulators of gene expression. Tissue-specific regulation occurs so that transcription of specific genes occurs only in certain tissues, while function-specific regulation also exists so that gene expression changes in reponse to a variety of biological challenges (Bondy and Roberts, 1969). As in the intact cell, when chromatin is tested *in vitro* for its capacity to support RNA synthesis, only a small fraction of the DNA is transcribed. The relative capacity of chromatin to support RNA synthesis *in vitro* is assumed to be representative of the relative transcriptional activity of the *in vivo* DNA (Bonner et al., 1968).

It seems highly likely that the multitude of morphological and metabolic effects which follow differential rearing are at least partially due to differential activation of the genome, so that the transcriptional activity of the genome to effect RNA synthesis is modified. It is clear that macromolecular metabolism is affected by experience (for reviews see

Glassman, 1969; Uphouse and MacInnes, 1974; Entingh et al., 1975), but in most studies the investigation of these metabolic changes has been measured several steps removed from the transcriptional process itself. As a result the contributions of transcriptional and post-transcriptional events could not be ascertained (Uphouse and Moore, 1978).

If gene activation is altered, then it might result from an altered rate of activity of already active genes, or the number and variety of active genes might change. Recent lines of study suggest that the number and variety of active genes may indeed be modified by environmental stimulation, but there is no evidence to delineate the specific RNA subtypes produced as a result of exposure to complexity and isolation. Alterations in brain RNA base ratios following training have been detected by several workers (Hyden and Lange, 1965), but the question of whether this indicates a preferential synthesis of certain types of RNA, novel RNA species (induction), or both, remains unresolved.

An attempt by Mushynski et al. (1973) to resolve this question for environmental complexity by means of competitive hybridization of hippocampal nuclear RNA was unsuccessful. No qualitative differences in RNA could be detected. However, this by no means rules out the possibility of qualitatively specific RNA synthesis, since the DNA:RNA hybridization techniques employed were sensitive only to highly reiterated gene sequences, that is, genes which occur in multiple copies. To overcome this problem Uphouse and Bonner (1975) studied RNA:DNA hybridization of rat brain RNA, with unique sequences of DNA (DNA sequences which were repeated only once to a few times per haploid genome). They reared enriched, isolated, and paired rats for 55 days, beginning at 25 days of age. At death the brains and livers were removed and RNA was extracted for hybridization with non-renatured DNA (unique single copy fractionated from rat liver).

The results indicated that the proportion of DNA hybridized by brain RNA from enriched animals was significantly greater than that from the nonenriched controls. Similar results have been noted by Grouse et al. (1978). In a further study in which RNA combined from both enriched and nonenriched groups was hybridized, a greater proportion of DNA was affected (i.e. additivity occurred) suggesting that there were some genes active in the nonenriched groups which were not active in the enriched. While this data is suggestive of heightened transcriptional activity following enrichment, the authors advocate caution in making this interpretation, since the differences detected lay near the limits of resolution of the technique.

Another measure of brain transcriptional activity can be obtained by determining _in vitro_ RNA synthesis by brain chromatin. Uphouse and Moore (1978) measured this in enriched, paired, and isolated rats differentially reared from 28 days of age for either 32 or 60 days. The activity in the whole brain was measured, and for the 60 day group the brain chromatin from enriched rats supported less RNA synthesis than did nonenriched. In a further study (Uphouse, 1978) the regional specificity of this response was determined, and it was found that in the cortex and cerebellum the enriched animals displayed greater synthesis, than did isolates or paired subjects. This regional specificity is of course consistent with previous morphological and biochemical studies which have generally found that enrichment causes heightened cortical and reduced subcortical values as compared with isolation. Further examination of the 60 day animals (Uphouse, 1978, personal communication) failed to detect significant differential activities in any of the regions examined. The relative template activitiy of whole brain also suggested that the 32 day enriched animals had lower values than nonenriched subjects but that the 60 day animals displayed no such effect.

Several measures clearly indicate that the sensory environment is capable of modifying DNA transcriptional activity of the central nervous system. This modification shows a regional specificity consistent with the patterning displayed by a variety of morphological and biochemical parameters. The time course of this activity also appears consistent with other findings, since for many parameters, peak values occur after about 30 days of differential rearing and decline thereafter. The absence of an effect at 60 days may indicate that genetic activation may be involved in initiation, but not necessarily maintenance, of structural and metabolic effects.

Protein

In contradistinction to the studies of simple sensory stimulation and deprivation (e.g. light versus dark), there has been relatively little investigation of protein changes following exposure to complexity-isolation. Bennett et al. (1964) reported no change in its cortical concentration per unit of wet or dry tissue. Nor was the concentration of hexokinase affected, though total activity followed the weight changes (Bennett et al., 1964; Krech et al., 1966).

Das and Altman (1966) have conducted an autoradiographic investigation of amino acid incorporation. Rats were differentially reared for 110 days and $H^3$ leucine was systematically injected one hour before death. Taken over all the regions examined, isolates showed a smaller uptake than did the complex subjects, which was attributed to the greater stress experienced by isolates. Unfortunately little can be made of this data, since only three pairs of subjects were used, no measures of stress were taken, and no information was provided concerning the stimulation experienced by each group in the interval between injection and death: Were they returned to their respective environments or exposed to identical ones? This is obviously a promising method warranting several further studies, but they will require several laborious extensions, including larger numbers of littermate subjects exposed in some cases to novel environments and stimulation. Also, in view of Rose's (1967) finding of multiphasic responses following exposure to light, they will have to be killed at various times after injection in order to allow temporal analysis.

A series of studies which is beyond many of these criticisms is that of Levitan et al. (1972a, 1972b, 1972c), who differentially reared female rats for 7, 16, or 35 days from weaning and then injected radioactive precursors intraventricularly two hours before death. During these two hours both complex and deprived rats were housed two to a cage, a technique which preliminary experiments indicated surprisingly enough gave the same results as when animals were returned to their home environments (Levitan, 1973, Personal Communication). The brain was then divided into cortical and subcortical samples although most of the subcortical radioactivity resided in the hippocampus. Perikaryal (cytosol, microsome, and mitochondria) and synaptosomal (synaptosol, mitochondria, and membrane) subcellular fractions were prepared and nuclear protein fractions extracted, and their protein specific radioactivity determined.

Within the hippocampus, incorporation into all subcellular fractions at seven days was 10-29% greater in enriched subjects, with these differences significant for all but mitochondria. After 16 days the differences in the synaptosomal fractions had increased 45-73%, while those of the perikaryal fractions had disappeared, and by 35 days no effects of rearing were detectable. Similarly, in the cortex all subfractions showed greater incorp oration at seven days and no perikaryal differences at 16 and 35 days. By contrast cortical synaptosomal incorporation was less at both 16 and 35 days.

For nuclear protein fractions only 16 and 35 day hippocampal samples were obtained and significant increases were found initially, but dissipated by 35 days.

Subsequent experiments were performed to determine some of the mechanisms mediating these changes. That they do not simply reflect differences in precursor pools was indicated by data suggesting that the size and specific radioactivity of the free leucine pool was equivalent in complex and deprived subjects. Differential stress appeared unlikely in view of the finding that corticosterone, which is the major adrenal steroid secreted by the rat, increased leucine incorporation into all fractions of the cortex, hippocampus and liver, whereas complexity had no such hepatic effect.

An important question related to the previously mentioned RNA specificity problem is that of the specificity of the new protein synthesis. A double labeling electrophoresis indicated that rather than being restricted to one or a few macromolecular species, the synthesis of all proteins appeared to be affected equally by environmental complexity, at least for hippocampal nuclei and synaptosomes.

The total amount of any labile substance in a biological system represents the net difference between synthesis and breakdown. Levitan et al.'s (1972b) next experiment suggests that both stages may be modified, at least for cortical synaptosomal proteins. In the only study to date of environmental effects on catabolism, rats were differentially reared for 31 days and injected with precursor at periods from 4 to 24 days before sacrifice. In the cortex but not hippocampus, significant turnover differences were noted for synaptosomal fractions; the protein from the enriched animals was the most stable. The half lives of the synaptosomal, mitochondrial and membrane proteins were approximately 18 days in the deprived rats, and 28 in the enriched; while for synaptosol the corresponding figures were 15 and 26 days respectively. However, no differences could be found in the perikaryal fractions.

Effects are largely transient, having disappeared by 35 days in all but the cortical synaptosomal fractions, where after the sixth day they are opposite to those of the hippocampus. This inverse cortical-hippocampal effect, with greater cortical persistence, is reminiscent of the reports of RNA (Glassman and Wilson, 1970) and protein (Hyden and Lange, 1972) precursor incorporation following various training procedures. It raises the question of the different roles of hippocampus and cortex in information processing, with the former being more concerned with consolidation of new information and the latter with long term storage. The partial transience of these effects parallels the time course of anatomical sequalae of differential rearing, and it would be extremely interesting to be able to localize these amino acid incorporation effects more precisely in order to see whether, like the anatomical, they display a marked regional specificity within the cortex (Rosenzweig et al., 1971) and hippocampus (Walsh et al., 1969). The selective effect on synatosomal components was considered by Levitan et al. to reflect the differences in the rate of axonal transport, rather than of protein synthesis, since all three synpatosomal subfractions were affected to about the same degree. It has been shown that the bulk of snyaptosomal proteins are supplied from the perikaryon by axoplasmic flow (Barondes, 1967).

At the same time as this series of studies makes a major contribution to our very limited knowledge of the short term biochemical processes elicited by complexity and impoverishment, it makes us even more aware of our ignorance. We know little of the precise relationship of these biochemical responses to the neural electrical activity which induced them or to the final long term anatomical and chemical effects which they mediate.

Until we know considerably more about the precise patterns of neural activity elicited by complex stimuli, it is unlikely that we will be able to interpret findings such as these adequately.

Glucosamine

Glycoproteins have been found to be an important component of synaptic membranes and may play a role in the acquisition of information by the nervous system (Bogoch, 1968; Brunngraber, 1969; Gombos et al., 1972). Since they had previously found differences in amino acid incorporation in nerve endings, and nuclear subfractions of both cerbral cortex and hippocampus following differential rearing, Ramirez et al. (1974) decided to repeat these previous studies using radioactive glucosamine to measure differences in glycoprotein metabolism. They therefore reared rats from weaning for 16 or 30 days in either enriched cages or in pairs in unfurnished cages. Following intraventricular injection of glucosamine, the animals were returned to their cages and killed one or three hours after injection. The cerebral cortices and hippocampi were dissected separately, pooled within each homologous group, and then fractionated into subcellular components, as were control samples of liver. After 16 days of differential rearing no significant environmental effects could be detected. After 30 days quite a different picture emerged. One hour after labeling, specific radioactivity was higher in enriched than in deprived rats, in both cortex and hippocampus. All synaptic components plus the nuclei and cytosol showed this trend, but due to the necessity of pooling samples it was not possible to test for significance levels.

After three hours labeling time, all components, with the exception of the hippocampal synaptic plasma membrane fraction, now displayed a reduced enrichment:deprivation ratio. This suggests that environmental enrichment increased the rates of uptake, incorporation, and disposal of the injected precursor. The authors suggest that this pattern is interesting in that critical regulatory phenomena are more likely to be dependent on rates of production than on total availability of some macromolecules. Only hippocampal synaptic plasma membrane of the enriched rats appeared to be accumulating more synthesized glycoproteins that its deprived counterpart, and it is tempting to relate this finding to synaptic consolidation in the hippocampus. Opposite metabolic patterns between hippocampus and cerebral cortex have been repeatedly noticed in this kind of study (Glassman and Wilson, 1970; Hyden and Lange, 1972; Levitan et al., 1972a, 1972b). This time course of the glucosamine results differs from those of protein synthesis (Levitan et al., 1972a, 1972b), in that the latter display significant effects after only 16 days exposure. They agree, however more closely with the results reported on changes in cortical depth and cerebral weight which, it will be remembered, display maximal effects after 30 days.

These studies provide the first glimmering of insight into the biochemical processes underlying the production of weight and anatomical changes following environmental complexity and deprivation. For protein at least it would seem that there is a net increase in synthesis and a partial decrease in breakdown following exposure to complexity, that these changes involve all protein classes rather than affecting specific subtypes, and that the effects display marked regional and temporal specificity.

SOCIAL AND NONSOCIAL STIMULI

The effects of either environmental complexity or deprivation cannot be determined alone, but only by comparison of differentially stimulated subjects. Where the aim is simply to identify effects of varying stimulation, it is logical to use extremes of stimulation. In this field of study the physiological extremes correspond approximately to the EC and IC conditions, and these have been the ones most widely used. However, when one wishes to identify the environmental factors and their precise attributes actually inducing changes, rather than merely identifying changes per se, one is then forced to compare groups of animals whose environments differ only with regard to the type of stimulation (environmental factor) under consideration. Obviously the EC and IC environments differ on an enormous number of stimulus dimensions, and the identification and quantification of their individual effects and interactions will be an equally enormous task. With rare exceptions, attempts to do so have been limited to comparing the effects of complexity and isolation with those of the standard colony situation.

The use by the Berkeley group of the standard colony social condition (SC) employing three subjects per cage as a baseline represents an arbitrary decision. For feral rodents it only remotely resembles the "natural" environment, and its "normality" for the laboratory rat is of course a moot point. Nevertheless, although arbitrary, it is useful, since it has served as the baseline for most animal studies in so many anatomical, physiological, biochemical and psychological fields. Its major disadvantage would seem to be that since the number of animals per cage differs from that in EC (usually 10-12), it is impossible to separate the relative contributions of social and physical stimuli in the EC condition. One way around this would be to use equal numbers of subjects in the EC and SC conditions.

Results available to date indicate a much more complex pattern of response to differential stimulation than anticipated. As might be expected, the effects of complexity and deprivation, as compared to the SC baseline, are usually opposite, but the magnitude and sometimes even the direction of the responses are dependent upon the genetic background and sex of the subject, the duration of differential rearing and the age at which it is initiated, and the brain region under investigation. The ultimate effect therefore probably represents a multiway interaction between all these factors and more. This of course makes the description inadequate to some degree, since the effects of multiple interactive factors can only be adequately and completely described in interactional terms, taking all factors into consideration simultaneously. But this renders a verbal nongraphic description hideously complex, and so for the purposes of this review, effects will be dealt with under individual headings.

## Brain Weights

At this point, it will suffice to simply compare the EC-SC and IC-SC differences derived from the standard experimental conditions of using male subjects differentially reared from weaning. It has been found in looking at brain weight from rats reared for 80 days, that weight is greater for EC than SC subjects throughout the cortex, 2.1% overall (Rosenzweig et al., 1971). IC is less than SC in all cortical regions except the ventral, where there is a small nonsignificant reversal, but the differences are small and amount to only 0.9% (NS) for total cortex. For the cerebral cortex, therefore, with the exception of the occipital region where the effects are approximately equal, the effect of superimposing complexity on the standard colony condition at first appears to be greater and opposite to the effect of isolation. In the subcortex, however, the opposite occurs, since while IC subcortical weight exceeds the SC by 2.1%, the EC and SC values are almost identical (Rosenzweig et al., 1971). Here then is evidence for an interaction effect between type of stimulation and brain region, indicating as might be expected that brain regions are differentially susceptible to the various stimulus components which distinguish the EC and IC conditions.

## The Influence Of Body Weight

With rare exceptions (Geller et al., 1976) it has repeatedly been found that isolates rapidly become heavier than EC subjects (e.g. Krech et al., 1966; Rosenzweig and Bennett, 1969; Ferchmin et al., 1970; Walsh et al., 1972b, 1973b). Rosenzweig et al. (1971) have found that the SC group tends to maintain an intermediate position for about the first 10 weeks. Thereafter its weight falls relative to the other groups, until at the termination of the 80 day period it is the lightest of the three. Since brain weight correlates with body weight, it is appropriate to use the latter as a covariate if one wishes to obtain some index of the effect of environment on brain weight independent of somatic mediation. Since cortex was heavier and subcortex lighter in EC as compared with IC subjects (4% and -1.1% of total cortex and subcortex respectively when averaged over all experiments, [Rosenzweig et al., 1971]),and since EC body weight was less, the effect of covariate adjustment was to yield increased EC brain values. The net effect of this general increase was therefore to further increase the positive EC-IC cortical weight difference and to reduce the negative subcortical effect, yielding for these experiments differences of 5.5% and 0.5% for cortex and subcortex respectively.

A similar effect was found by Walsh et al. (1974) with animals differentially reared for periods of from 18 to 530 days. Overall, forebrain and wholebrain weights were 3.4% and 2.6% greater and body weight was 7.6% less in EC than in IC subjects, so that the effect of covariation was to increase forebrain and wholebrain differences to 4.5% and 3.6%. It is therefore obvious that the observed EC-IC difference represents the net effect of two processes: the first a direct environmental influence on the brain, and the second an effect mediated by somatic influences.

The effects of covariate adjustment are even more marked if one wishes to compare EC-SC and IC-SC differences, (Rosenzweig et al., 1971). Since the EC-SC terminal body weight difference was relatively small, the effect of covariate adjustment was also small; but as the IC-SC body weight difference was the largest of all, the corresponding covariate adjustment was also large. Since isolates were somatically heaviest, the effect of covariate adjustment was to reduce all isolate brain weights. Since the isolate cortex is the lightest to start with, the effect of adjustment was to further reduce it (from -0.9% to -5.4%) to the point where the negative IC-SC difference was greater than the positive EC-SC difference. For subcortex, however, the opposite occurred. Since the IC subcortex was originally the heaviest, adjustment reduced the IC-SC difference from positive (2.1%) to negative (-2.8%).

The C/S ratio is of course unaffected by body weight, which is one of the reasons for its comparative stability when compared with either of its components alone, and thus is effectively unaltered by covariate adjustment (Rosenzweig et al., 1971). For similar resons the forebrain: hindbrain ratio is not susceptible to covariation (Walsh et al., 1974).

Whereas initially it appeared that the cerebral weight and enzyme effects of superimposing enrichment on an SC baseline were greater than those of isolation, it appears that the reverse is true when the effects of body weight are taken into account.

## AChE and ChE

For both enzymes the IC-SC difference in activity per unit weight was found to exceed the EC-SC throughout cortex and subcortex (Rosenzweig it al., 1971). Since the activity per unit weight measures do not correlate with body weight, they do not require covariate adjustment. The original values for enzymic activity may therefore be compared with the adjusted ones for weight, and together would seem to indicate that compared with the SC baseline and independent of somatic modification, the effect of isolation is greater than that of enrichment, especially for AChE and ChE activities, where enrichment is almost without effect.

## Cortical Depth

As yet no details of EC-SC-IC effects of 80 days differential rearing on cortical depth are available, but Diamond et al. (1972) have presented a number of 30 day studies. The reader should remember that these findings are for males only, and that females may display somewhat different effects (e.g. for the C/S ratio, larger enrichment effects). As might be anticipated, the effects are in some ways analogous to those of weight, but the greater degree of localization conferred by depth studies reveals a more intricate pattern of regional response specificity. When differential rearing commences at 25 days of age, in the medial cortex both the EC and IC subjects differ almost equally from the SC. However, in passing laterally the EC effect becomes less and less marked while that of isolation remains approximately constant. Since no covariate adjustment was attempted in this study, it is impossible to determine to what extent the effects are independent of somatic mediation.

To what extent does the SC-IC difference truly reflect social factors, and to what extent is it due to the different cage sizes? A number of lines of evidence suggest that the interaction between the two factors is synergistic. Isolates in large cages or even EC cages show almost no EC cerebral effects (unless primed with a stimulatn drug). Increasing group size tends to enhance cerebral measures in the EC direction, but these do not attain EC levels, at least for groups numbering up to 12 (Rosenzweig and Bennett, 1977).

One investigation of the influence of various types of inanimate environments has been conducted by Rosenzweig and Bennett (1978). They compared standard EC and IC conditions with those comprised by using bare cages, a cage containing a simple maze in which barriers remained constant and which animals had to transverse to obtain nourishment, a complex maze in which the pattern of barriers was changed daily, and a semi-natural outdoor condition of presumably greater stimulus variety and complexity than EC. All experimental treatment, except IC, involved equal numbers of animals per cage so that the effects of the inanimate physical environment could be determined precisely.

In general both mazes elicited similar though smaller effects than EC. Group living also produced similar effects but these were small and less than those for the mazes. On the other hand, the impact of the semi-natural outdoor environment was clearly and consistently greater than EC, though once again the regional and parameter patterning was similar. This study clearly demonstrates that the addition of complex inanimate stimuli produces effects which are greater than those of social grouping alone. Furthermore, a variety of physical environments and tasks are effective and qualitatively similar, though they may vary in the magnitude of the responses they elicit. It seems likely that further studies will support these principles and may also detect qualitative differences in the neural consequences of differing physical stimuli.

For each of the parameters examined to date the effects of complexity and isolation, or of animate and inanimate stimuli, cannot be considered as qualitatively or quantitatively equivalent, and each displays a distinct pattern of regional response specificity. This probably represents an example of what may one day come to be recognized as a general law of environmental sensory effects on brain: that every environment or stimulus complex is associated with its own unique pattern of brain response; a kind of "psychophysical equivalence". It is currently impossible to interpret adequately the functional significance of the differential patterns elicited by environments such as EC and IC, which differ on such an enormous number of stimulus dimensions.

## Super EC

One question of obvious interest is just how far cerebral values can be raised. One obvious line of approach to this question is to attempt to further increase the complexity of rearing environments, and three research teams have now done this. Kuenzle and Knusel (1974) designed a "super-enriched" environment involving groups of 70 rats housed in two large interconnected cages which required them to shuttle across a bridge with changing gates and signals in order to obtain food and water. The super-enriched anumals were found to exceed the regular EC rats, in the weight and ChE:AChE ratio in occipital cortex and length of the cerebral hemispheres, but not in RNA:DNA in occipital cortex. Ferchmin et al. (1970) employed larger cages and rotated animals between cages several times per day. They found a more marked early increase in forebrain RNA than that displayed by the typical EC animals. Then again Rosenzweig et al. (1972) have found that rats reared in an outdoor semi-natural environment display greater effects. Thus it is clear that the typical EC condition does not represent the upper limit of cerebral development, which remains to be discovered.

## Locomotion

One of the mechanisms which is most frequently advanced as a possible factor involved in the production of environmental effects is differential locomotion. On first sight this seems a plausible hypothesis, since the complexity rats are at times highly active, exploring toys, and interacting with one another, while isolates are confined to a smaller cage in which considerably less locomotion is possible. However, a number of experiments involving controlled differential locomotion have failed to produce any evidence in support of this. Several studies have employed activity wheels in which animals can exercise at will (Krech et al. 1960; Rosenzweig et al., 1968; Huntley and Newton, 1972), while Ferchmin and Eterovic (1977) trained rats to walk upside down on a wire mesh ceiling and to pass barriers in order to get food. Furthermore, in experiments on formal maze training, rats running in a two-meter-long alley some 50 or more times per day, over a 30-day period, displayed only minimal effects on cerebral measures (Rosenzweig and Bennett, 1977).

## Handling

Similarly, it seems clear that differential handling cannot account for the neural effects. Thirty or 60 days of handling for two minutes each day had no effect (Krech et al., 1960), while a combination of locomotion plus handling was ineffective, except perhaps slightly in the somesthetic region of the cortex (Rosenzweig et al., 1968). Rosenzweig and Bennett (1977) concluded that neither postweaning handling, nor differential locomotion, nor the combination of these can account for the EC-IC brain differences.

## TEMPORAL FACTORS

Considerable effort has gone into determining the temporal patterns of brain changes induced by complexity and isolation. Both the duration of differential rearing and the age at which it is initiated have been shown to be important determinants of the magnitude, and possibly even the direction, of induced effects. Additional studies have examined the effects of varying the duration of exposure each day, and others have looked at their permanence and reversibility.

### Duration Of Differential Rearing

Initially it was assumed that increasing duration would merely increase EC-IC differences to asymptotic levels. However, temporal analysis has revealed that in point of fact several effects show transitory components such that when separation is continued beyond a certain period differences diminish. Furthermore, different parameters and brain regions display different temporal patterns, effectively indicating an interaction between duration, parameter and region.

#### *Weight*

Because of the initial assumption that prolonged periods would be necessary to elicit detectable effects, most studies have employed durations of 15 to 160 dyas. However, recent work has discerned effects after as little as four days and in the Berkeley lab the following patterns have emerged (Rosenzweig et al., 1967; Rosenzweig and Bennett, 1978). After 15 days from weaning, effects were greatest in the somesthetic region and, surprisingly, smallest in the occipital (Fig. 4). Not unexpectedly, as the duration was increased from 15 to 30 days, the

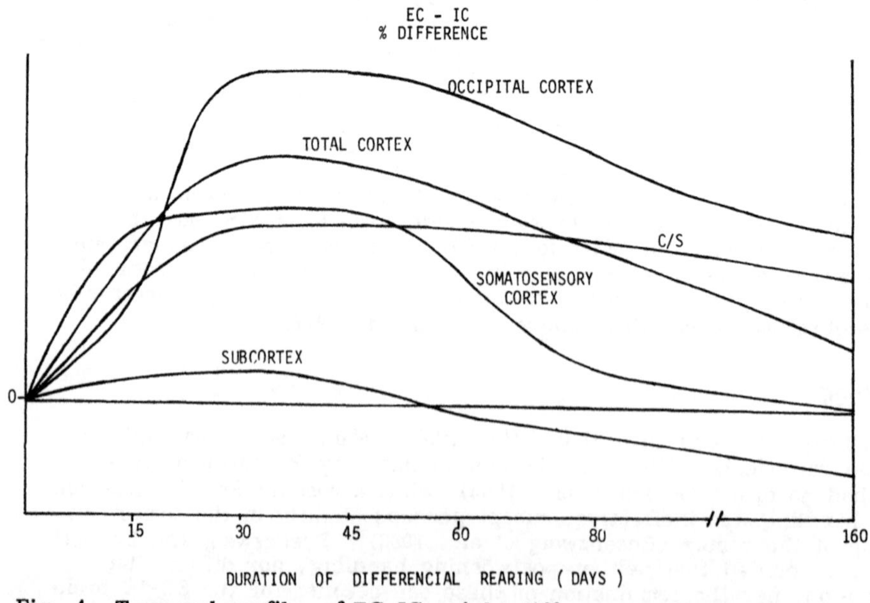


Fig. 4. Temporal profiles of EC-IC weight differences. Drawn from data of Rosenzweig et al. (1967a; 1971).

magnitude of the EC-IC differences increased on almost every measure (that for the somatosensory sample being the exception), with occipital now showing the greatest effect. As the duration was prolonged, however, from 30 to 45 days, the differences began to diminish, with the exception of the somatosensory cortex and the C/S ratio. As the duration was further extended through 60, 80 and 160 days, most differences continued to decrease, this time with the exception of the occipital sample and the C/S ratio; so that by 160 days, the only residual cortical effect was in the occipital region. The net effect for total cortex was a maximum (5.9%) at 30 days, followed by a steady decline thereafter until at 160 days only a 2.5% residue, due largely to the occipital region, remained. The temporal pattern of subcortex was similar, but the magnitude was considerably less, reaching only 0.5% at 30 days and declining steadily thereafter to reach negative values within 80 days. This parallel between the temporal patterns of the total cortex and subcortex explains the temporal stability of the C/S ratio. It will be seen subsequently that the negative EC-IC effects noted in the subcortex are largely the result of the greater body weight of the isolates.

The regional specificity of the temporal patterns of cortical change is best exemplified by direct comparison of the various areas (Fig. 5). The somatosensory change reached a maximum within 15 days, at which time it was almost twice that of the occipital. However, these somatosensory effects did not grow larger with prolonged exposure, but rather remained steady until 30 or 45 days and then dropped sharply. By comparison, the occipital differences continued to increase until 30 days and showed little tendency to decline thereafter, in contradistinction to the ventral and remaining dorsal cortices, which also showed increasing effects through 30 days but subsequently dissipated.

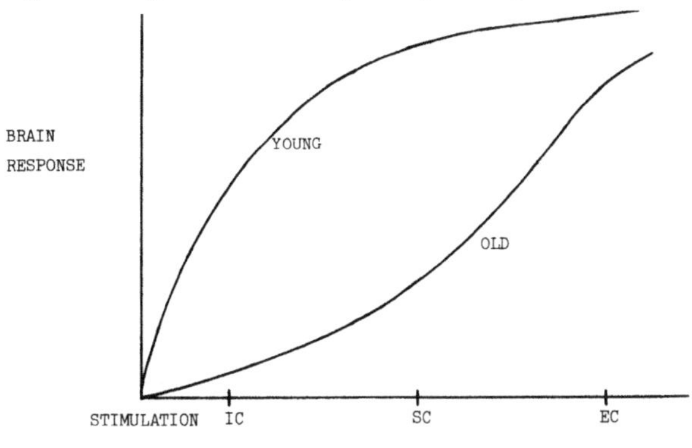


Fig. 5. Hypothesized relationship between the magnitude of brain response and the level of stimulation, as a function of age. With increasing age the stimulus range within which maximal brain responsivity occurs (shown by greatest slope of the line) occurs at higher levels of stimulation.

The only other temporal analysis utilizing the standard EC-IC conditions is that of Walsh et al. (1974), who exposed Wistar rats for periods ranging from 18 to 530 days and measured whole, fore, and hindbrain weights as well as the derived measure of forebrain:hindbrain ratio. Hindbrain showed no effect, but the other parameters had attained significant values by 18 days (5.2%, 6.1% and 4.7% for wholebrain, forebrain and forebrain:hindbrain ratio respectively). Similarly to the patterns reported by the Berkeley group, effects tended to remain significant throughout 40, 60, 80, 90 and 120 day periods, but to have largely disappeared by 160 days.

An increase in the complexity of the EC condition was developed by Ferchmin et al. (1970). They employed very large (approximately 30 cubic feet) EC cages and changed animals from one cage to another several times a day, a maneuver which they felt was more effective in eliciting behavioral activity than simply changing toys. At the beginning of experiments they added several older "leader" rats who had previous EC experience, and with the example of the leaders, new animals began to explore and climb earlier than they otherwise would have. The brain weight measures used were similar to those of Walsh et al. (1974), namely forebrain, hindbrain and wholebrain, and hindbrain was similarly found to be unaffected. The temporal analysis extended from 2 to 60 days of differential rearing from weaning. The forebrain exhibited a significant 2.0% modification after only four days and a 1.5% value after another four. Whether this represents a significant decrease is not mentioned. Thereafter, the forebrain value rose to 5% at 30 days and to 4.5% at 60 days. The RNA concentration followed a very similar pattern, rising to a maximum of 4% at four days, falling to a minimum of 0.5% by eight and thereafter showing a slight increase to about 1% at 30 and 60 days. Thus there would appear to be a short initial four-day phase of rapid development, followed by a more gradual phase within the first 30 days. In general the RNA:DNA ratio rises more rapidly than weight, remains larger during the first 30 days, and then becomes smaller. This measure therefore appears to be a rapid and sensitive index, as might be expected if it represents the cellular response of producing new RNA as the initial step in a biosynthetic response to stimulation (Rosenzweig and Bennett, 1978).

## *The Geriatric Brain*

Almost without exception studies of environmental stimulation have focused on the young animal and have employed short exposure periods by comparison with the subject's life span. In part, this reflects the gradual retreat from the concept of brain immutability, since it has been widely assumed that cerebral plasticity is restricted to the young developing animal and hence that changes would not be retained as animals age and would not occur de novo in the geriatric brain. Support for this is available from the finding that the anatomical and chemical effects of exposure to complexity tend to diminish after about 30 days (Rosenzweig et al., 1967, 1971).

These assumptions of limited or nonexistent geriatric cerebral plasticity are obviously very important ones. A study by Cummins et al. (1973) aimed at testing them: firstly, by determining whether the effects of differential stimulation might not be detectable after considerably longer periods than have previously been employed, and secondly by adding a further form of stimulation in old age, determining whether evidence for de novo cerebral change could be detected in the geriatric brain. The experiment was also designed to determine whether the nature and magnitude of the response to stimulation in old age was dependent on the animal's prior experience. Differential rearing continued uninterrupted for 509 days from weaning, after which half of each group were removed from their respective environments for 15-30 minutes each day for 21 days of Hebb-Williams maze testing. By the end of the experiment the 555-day-old animals had lived for approximately two thirds of their maximum anticipated life span.

Complexity-exposed subjects made significantly fewer maze errors than isolates, and both environmental complexity and behavioral testing augmented cerebral weights and dimensions, although the effects of testing were evident only in isolates. Complexity subjects displayed a 7% greater forebrain and hence wholebrain (4.6%) weight than their untested isolate counterparts, while hindbrain showed a nonsignificant trend in the opposite direction, thus yielding a large 8.4% forebrain:hindbrain ratio effect. Cerebral length but not width was also greater in complexity than in untested isolate subjects, and as has been found with younger animals (Walsh et al., 1971a, 1973a), the product of length and width yielded a more sensitive (4%) index than did either of the primary measures separately. Behavioral testing also acted to increase forebrain (5.2%) and wholebrain (4.3%) weight as well as cerebral length, but the effects were evident only in isolates, thus yielding a significant testing X environment interaction. Testing effects were not as clearly localized to the forebrain, so that in contradistinction to complexity-isolation, the forebrain:hindbrain ratio did not prove a particularly sensitive index of this kind of stimulation.

These effects of environmental complexity on behavior, brain weight and cerebral dimensions resemble those previously identified in younger subjects (Cummins et al 1971; Sherlock et al., 1973; Walsh et al., 1971a, 1971b, 1973a). If the effects of testing are ignored, then their magnitude is as great as any found in this laboratory. This finding is particularly remarkable when we remember that complexity-isolation differences tend to diminish after 30 days and are only very slightly in evidence by 160 days (Rosenzweig et al., 1971; Walsh et al., 1974). Although it has been widely assumed that the effects of differential rearing continue to diminish asymptotically, it is now evident that a marked recrudescence may occur with very prolonged treatment.

The finding of significant cerebral morphological effects following maze testing is particularly noteworthy. Similar procedures have been found to be relatively ineffective in younger animals reared under normal laboratory conditions (Rosenzweig et al., 1971), although Diamond (1976) has noted some effects on cortical depth and Greenough (1976) on peripheral dendrites.

Since in this study the effects of testing were evident only in isolates, the question arises as to whether the prolonged period of isolation and stimulus deprivation served to sensitize these animals. Certainly the isolates showed a marked aversion to handling and seemed generally to display a higher level of arousal than their complex littermates. Such an observation offers support for the hypothesis that one of the major mechanisms mediating the induction of environmentally induced brain changes is the arousal reaction. The absence of a significant cerebral response to testing in complexity reared subjects might be seen as due to a lesser sensitivity of the mechanisms mediating the production of anatomical change, or as an example of the existence of a ceiling effect such that cerebral metabolism was unable to sustain a further increase above that already induced by complexity. Whatever the mechanisms involved, the finding is interesting in demonstrating the dependence of the nature of the cerebral response on the previous experiential history of the animal. This has been widely reported for behavior (e.g. Denenberg, 1969a; Bernstein, 1971, 1972; Sherlock et al., 1973) and supports the concept of environmental therapy or enrichment ot overcome the effects of prior deprivation: e.g. the so-called cultural deprivation syndrome (for reviews see Jessor and Richardson [1968] and Birren and Hess [1968]).

Five-hundred and fifty days is certainly aged by comparison with most laboratory rats, but some researchers questioned whether it truly represented a geriatric stage and, if not, whether environmental effects would still be found. To answer this question Walsh and Cummins (1977) reared 20 triplet sets of Wistar rats from weaning in EC, SC and IC for 900 days. At the end of this period the animals had suffered 50% mortality and the remainder appeared truly geriatric. One third of the survivors were killed, and the remainder were all placed in EC for further periods of 16 or 36 days.

The effect of 900 days of differential rearing was to increase the weight of the forebrain and hence wholebrain in social and complexity animals (which did not differ from each other) relative to the isolates. The hindbrain exhibited a nonsignificant trend in the opposite direction, thus yielding a large forebrain: hindbrain ratio effect. This pattern is similar to that found in younger animals. Placing the social group in complexity elicited no detectable effect, but for isolates there was a striking and significant increase in forebrain weight and forebrain:hindbrain ratio such that by the end of the 36 days the effects of the prior 900 days had been largely overcome.

It is obvious that, contrary to previous thinking, the mammalian brain may not only maintain structural and behavioral changes originally elicited in youth, but may also retain a considerable degree of cerebral plasticity in extreme old age, and the effects of previous experiential deprivation may be largely reversible. Such findings would seem to hold widespread theoretical and practical implications and to suggest that the adaptive potential of the geriatric brain may have been considerably underestimated.

*Cerebral Length*

Cerebral length continues to grow in the rat, rapidly for the first 20 days after birth and more slowly thereafter. Few temporal analyses of EC-IC effects on cerebral dimensions are available, but the limited data suggests a slow onset followed by a relatively stable effect throughout the major portion of the life span (Walsh et al., 1971a, 1973a; Cummins et al., 1973; Cummins, 1978). Thus the thirty day cerebral length effects were 1.2%, but subsequently stabilized around 2%. The phenonemon is somewhat more complex than it first appears since Cummins (1978) has recently found that the EC-IC difference is initially most marked in smaller brains, but with longer duration this effect reverses. This finding is examined further in the discussion of the rank order effect. Replications of these findings will of course be necessary to demonstrate their reliability and significance; if validated they will prove extremely interesting since they stand in such contrast to those of cortical weight and depth.

*Cortical Depth*

As yet there have been few studies allowing a temporal analysis of cortical depths, but the information currently available would seem to suggest that, as might be expected, depth and weight display similar patterns. From the experiments of Diamond et al. (1971, 1972), information is available on sections corresponding to occipital, somatosensory and frontal cortex for 30, 56, and 80 days of differential weaning. While direct statistical comparision has apparently not yet been performed, it would seem that all regions display maximum EC-IC differences within 30 days and decrease thereafter. This decrease is least in the occipital area, but by 80 days, differences have largely disappeared within all other regions except the frontal (Walsh et al., 1972).

Diamond (1967) has discussed two hypotheses to account for these regional patterns of weight and depth change. The first is that vision plays a predominant role in the rat's adaptive behavior. However, behavioral studies make this seem most unlikely, and it has been reported that the visual cortex of blinded animals shows the typical EC response

(Krech et al., 1963). The second hypothesis is that the visual cortex is more plastic than other cortical regions; this is supported by the finding that this is one of the last cortical regions to myelinate (Jacobson, 1963). Such an hypothesis appears unwarranted in view of the fact that initially cortical areas all increase to approximately the same degree, differing mainly in the time course of these changes. Once again the basic question becomes one of exploring the different time courses of the changes, and the relative absence of transient components in the occipital region. A possibility which has not yet been mentioned but which must be kept in mind is that the occipital cortex does exhibit transient changes but that these are of extreme duration, and have thus not yet been identified. If this is so, one is really no better off, since it merely reaises the question of why this region shows such a markedly different time course as compared with other areas.

The finding of typical EC changes in blinded animals argues strongly for nonvisual functions of the occipital cortex, and for EC changes not being mediated solely by visual stimuli. Lashley (1943) first suggested that the striate cortex mediated nonvisual functions when he found that maze learning in the rat was more severly disrupted in subjects that had sustained both striate cortical ablation and peripheral blinding, than in peripherally blinded animals. It is now know that the striate cortex responds to nonvisual stimuli (Lubar et al., 1967) and contains units responsive to somesthetic (Horn, 1965), auditory (Jung et al., 1963) and noxious (Kimble, 1963) stimuli, but precisely which or all of these stimuli or their attributes actually induce EC changes, remain unknown.

Another hypothesis which would appear to account for the occurence of effects in blind animals, and the transience except in the occipital region, is that the basic underlying neurophysiological mechanism mediating enrichment effects is the arousal reaction (see Chapter 8).

## *AChe And ChE*

The independence of these enzymes is very clearly displayed by their markedly different temporal patterns. Like weight, changes in total AChE activity reach a maximum within 30 days in both cortex and subcortex and thereafter decline (Rosenzweig et al., 1967). Since the cortical increases in total activity are less than those of weight, the EC-IC differences in AChE/weight are negative, and like total activity are maximal within 30 days and decrease thereafter.

ChE on the other hand displays a slow build up such that EC-IC differences in ChE/weight at 30 days have attained only approximately half of their 80 day readings in both cortex and subcortex (Rosenzweig et al., 1971). More prolonged studies will be necessary to determine the question of asymptotic levels and transitory components. If, as has been suggested, ChE may be taken as an index of glial numbers and activity, then perhaps the development of glial changes follows a similarly slow progressive buildup. Unfortunately very little in the way of temporal analyses of cell counts and histological parameters other than cortical depth is available.

## Age At Initiation Of Differential Rearing

A question of obvious interest, and one which increases in importance as experimenters age, is whether the brain is differentially sensitive to stimulation at different ages. Using a different terminology this may be rephrased as, do "critical" or "sensitive" periods exist? Research on the effects of handling the neonate rat (for reviews see Denenberg et al.,

1968; Walsh and Cummins, 1976b) suggest that they might, but the evidence for environmental enrichment and deprivation indicates that the effects of varying the age of initiation are qualitatively and quantitatively considerably less marked than for those of neonate handling. Initial ages have varied from six to 310 days and in each case similar, though not identical, changes have been observed. Most studies have employed postweaning initiation ages and these will be dealt with first.

*Brain Weight*

Subjects differentially reared for 30 days beginning at 60 (Rosenzweig et al., 1971) or 290 days (Riege, 1971) display a slightly less marked EC-IC effect in occipital and total cortex than do those commenced at weaning. It might be, however, that this is due to a slower development of these differences than in the younger animals, since in subjects treated for 90 days from 290 days of age (Riege, 1971) the effects were at least as marked as in their younger counterparts (Rosenzweig et al., 1971). Interestingly enough, 210 or 240 day old rats do not appear to exhibit the enhancement of EC effects which amphetamines appear to elicit in weaning subjects, perhaps because a heightening of exploration is not elicited in the older animals (Bennett et al., 1973).

When commenced at weaning, the effect of differential rearing is to increase the weight of the isolate subcortex above that of the EC; this difference is due to the greater isolate body weight. Rosenzweig et al. (1971) have found that when differential rearing is commenced at day 60 or after, this body weight effect does not occur, and hence EC-IC subcortex weight differences are positive. Accordingly the C/S ratio is slightly reduced by late initiation. However, this failure of differential body weight development is not invariant since in another lab two out of five such experiments yielded significant effects (Cummins and Walsh, 1978).

Initiation after weaning also modifies the relative EC-SC and IC-SC effects. Following 30 days of differential rearing from weaning (25-55 days), and after taking body weight differences into account, the major component of the EC-IC cortical difference is due to isolation rather than enrichment, whereas the reverse is true when commenced at 60 days (Diamond et al., 1972). The exact magnitude of the component contributions varies from area to area, indicating an interaction between age of initiation, type of environmental stimulation and regional specificity.

*Cortical Depth*

The greater spatial resolution afforded by cortical depth measures allows a more precise definition of the regional specificity characteristics of this interaction. In the 25-55 day animal enrichment makes a contribution equal to that of deprivation only in the medial occipital area, but peters out as one moves either laterally or longitudinally (Diamond et al., 1972). And in the extreme lateral sections, the EC-SC depth effect actually shows a slight negative trend.

In the 60-90 day subjects (i.e., differentially reared for 30 days from age 60 days), as with cortical weight, the enrichment contribution to the EC-IC depth effect is more marked. As with the 25-55 day group the major enrichment depth effect, though larger, is still most prominent in the medial occipital areas, with the more anterior somesthetic and frontal regions being less susceptible to enrichment, though still showing isolation effects by comparison with the social control baseline. Once again the enrichment effect also decreases laterally, but this effect is not quite as marked as in the younger animals. Cortical depth studies therefore confirm and extend the cortical weight findings of an interaction between regional specificty, age of initiation and type of stimulation. Perhaps at least one further factor must be added to this equation, since Diamond et al. (1972) could find no significant age of initiation effect on the magnitude of the EC-IC depth effect in subjects differentially reared for 80 days, when rearing was started at either weaning or 105 days. Thus the duration of differential rearing may be an inherent component of a more complex interaction.

*AChE And ChE*

As with weight and depth, the esterases also respond to a later age of initiation with an increase in the relative effect of enrichment as compared with isolation. In the standard 80 day experiment commenced at weaning, most of the EC-IC differences in both ACh and ChE are due to impoverishment, rather than to enrichment, when compared with the SC baseline. However, when differential rearing begins at 105 days of age, this effect is reversed for AChE as regards both AChE/weight and total AChE activity (i.e., enrichment is effective but impoverishment is not). This age of initiation effect is less marked for ChE, since impoverishment still remains effective while enrichment is effective for total ChE activity but not for ChE/weight (Rosenzweig et al., 1971). The change in total ChE is no longer greater than that for weight.

It would therefore seem that the "critical period" concept does not apply to environmental complexity, but that delaying the age of initiation beyond weaning results in a less marked impoverishment and a more marked enrichment effect. One explanation for this finding might be that as animals mature they are better able to process greater amounts and complexity of stimulation, so that great input is required to elicit equivalent neural effects. This would also be consistent with the organism's greater experience such that fewer stimuli present novel demands. Looked at from the opposite perspective maturation confers some degree of immunity against deprivation. However, the relative increase in the enrichment, compared to the deprivation, response depends on the duration of stimulation and displays a marked regional and parameter specificity. There is therefore evidence for a multiway interactive determination of brain responses, which must be even further expanded to include the genetic background and sex of the subject. (See Fig. 5, p. 55.)

*Preweaning Enrichment*

Temporal analysis has been extended to the preweaning age range by Malkasian and Diamond (1971). Rat pups were differentially reared from day six for 8, 13, and 22 days (i.e., until 14, 19 and 28 days of age), in litters of three with their mothers in one of three ways:

1. A single litter and mother in standard colony cage (unifamily environment, UFE);
2. Three mothers and their litters in a large cage (multifamily environment, MFE); and
3. Three mothers and their litters in a large cage together with toys (enriched condition-multifamily environment, EC-MFE).

Cortical depth did not differ between UFE and MFE in the occipital and somatosensory regions. However, UFE and EC-MFE cortices differed significantly at all three ages in both areas. The onset was slowest in the medial occipital region, where differences did not reach significant levels at 14 days, whereas the effect was most marked in the lateral occipital area, reaching 16%. The magnitude of 16% is considerably greater than the usual adult range of 5-10%, suggesting a greater plasticity in infancy. This is in contradistinction to the postweaning situation where the greatest EC-SC cortical depth difference occurs medially, at least at 30 days, the shortest duration for which histometric data is available. This disparity suggest either an age of initiation effect, or that depth changes in the medial occipital cortex are of less rapid onset but continue to develop over a longer period than those which develop more laterally. Malkasian and Diamond suggest that the delayed differentiation in this area, a primary visual center (Kreig, 1946), may be due to an absence of visual input (the eyes of the rat pup do not open until the second week), coupled with the relative dearth of other input because of the undeveloped state of associative interconnections. A growth of neuropil was evident in the lateral geniculate of the 28 EC-MFE animals, which would also seem to suggest a partial visual mediation of cortical effects.

Neuronal perikaryon and nuclear areas were measured in layers II and III of the somatosensory cortex at 28 days. Differences between EC-MFE and UFE subjects amounted to 19% and 16% for perikarya and nuclei respectively, while neuronal density per unit volume of tissue was reduced by 17% in the EC-MFE. These figures are comparable to, though larger than, those noted in the standard 80 day experiments (Diamond, 1967). A similar study of RNA:DNA ratio found that two weeks of preweaning treatment resulted in a slightly smaller effect than postweaning on this parameter (Rosenzweig and Bennett, 1978). These findings extend the generality of the cortical histometric and chemical data to the infant animal.

### Partial Daily Exposure

In the same way that it was originally believed that the more prolonged the total duration of differential rearing, the greater would be the cerebral effects, it was also felt that the greater the period of exposure per day, the greater would be the resultant change. However, with the findings that shorter durations of differential rearing were effective and that formal maze training added very little, it was wondered whether periods of less than 24 hours per day might also be effective. In fact it was found that two hours per day over a 30 day period was sufficient to elicit full weight, AChE, and ChE effects (Rosenzweig et al., 1968), except in inbred Fisher strain rats and that as little as 30 minutes per day elicited significant though not complete effects (Bennett et al., 1973). It may be that shortening the period of daily exposure reduces the speed with which effects develop, since with only two hours daily exposure no significant effects could be detected after only 15 days. Future studies may perhaps reveal the precise functional relationship between the amount of brain modification and the amount and distribution of environmental exposure. This finding fits well with behavioral studies which show that animals are maximally aroused and exploring within the first few minutes of daily exposure to EC or new toys, and that this activity diminishes markedly over the next hour (Ferchmin and Eterovic, 1973; Walsh and Cummins, 1976b; Cummins et al., 1978).

## OTHER INDEPENDENT VARIABLES

For the sake of simplicity, experimenters usually isolate individual independent variables and view their effects as being additive. In point of fact, any behavioral or neural end state probably represents the outcome of a complex interaction, rather than simple addition, between all genetic and experiential factors (Walsh and Cummins 1976b, 1978; Walsh and Shapiro, 1978a). In view of this interactional determination, it is impossible to predict in any simple additive way the effects of varying other factors in addition to environmental complexity. With an infinite number of additional variables that one might wish to manipulate this potentially makes for an extremely complex situation. Yet only a few factors, particularly excitant and depressant drugs and the sex and genetic background of the subjects have been examined. There will probably be an explosion of experiments in this area as the effects of more and more factors are determined. Some of these (e.g., malnutrition, hypoxia, brain trauma, and hormones) are of direct clinical relevance; hopefully the study of environmental impact on their effects might lead to new modes of therapy (see Walsh and Greenough, 1976a, 1976b and Chapter 7).

## Injections

Before we examine the effects of drugs, it is appropriate to note that the injection procedure itself may be capable of modifying the environmental effects. Eterovic and Ferchmin (1974) observed that a single intraventaneous water or intracranial saline injection dissipated the effect of six days EC-IC rearing on forebrain weight. The early handling literature clearly demonstrated the widespread and long lasting effects of handling neonates, so it is perhaps less surprising that injection plus the attendant handling procedures can affect brain structure even after weaning. In any case the potency of experimental procedure such as injections must be taken into account in future research.

## Drugs

It is a long known but little used fact that CNS stimulant and depressant drugs are capable of modifying the rate of recovery from brain lesions (Ward and Kennard, 1942; Watson and Kennard, 1945). This led the Berkeley group to question whether such drugs might not modify the effects of environmental stimulation. Their hypothesis has now been confirmed by a series of experiments in which subjects were injected with either drug or saline each day before being placed in differential environments (Rosenzweig and Bennett, 1968; Rosenzweig et al., 1971; Bennett et al., 1973). On exposure to EC, animals initially actively explored and engaged in social interaction, but thereafter habituated and by the end of two hours, about a quarter of them were asleep. A similar behavioral pattern is apparent when new toys are added to the EC cage each day (Walsh and Cummins, 1975). The effect of phenobarbitol was to increase the rate of habituation so that while during the initial half hour phenobarb and saline injected subjects appeared to be equally active, by the end of two hours about three quarters of the phenobarb subjects were asleep. With methamphetamine, on the other hand, young weanling subjects explored more and engaged in less social interaction; this heightened exploration and motor activity was maintained throughout the two hour testing period. Interestingly enough, this did not hold for older (210-240 day) subjects; compared with saline controls they remained more awake and showed severe startle reactions to minor stimuli. On the other hand they engaged in less exploration and far more social interaction, particularly that which involved the assumption of sparring positions, rearing up on their hind legs and facing each other as though preparing to fight. These behaviors are interesting both in their own right and because they afford some insight into the concomitant brain changes.

Methamphetamine

The effect of methamphetamine when given to weanling animals was to significantly augment the brain weight effects of environmental complexity while exerting a less marked influence on the cholinesterase enzymes. An interaction between environment and drug was apparent, since effects of methamphetamine were evident only in the EC subjects and not in the isolates, except for a small increase in dorsal cortex weight. Thus in subjects differentially exposed for two hours per day for 30 days, methamphetamine-treated EC subjects exhibited significantly heavier brains than their saline counterparts, and the regional patterning of weight increases was similar to that elicited by EC itself. In the isolation condition, on the other hand, drug and saline subjects had almost identical values. This suggests that the mode of action of amphetamine is largely one of augmenting the effects of environmental stimulation. For AChE and ChE, on the other hand, the trend was towards an EC like effect in both EC and IC conditions, although in neither environment was this significant. This enhancement of EC brain weight effects was also seen in experiments in which the amount of differential exposure was reduced to either 30 minutes per day or only 15 days. It will be remembered that the behavioral effects of methamphetamine were notable in older subjects who did not exhibit the increases in exploration which characterized the younger animals. It is therefore interesting that the older animals did not show any significant brain weight effects of the drug, suggesting that active exploration is an essential factor.

In an effort to further define the environmental conditions necessary to elicit cerebral effects, Rosenzweig and Bennett (1972) placed rats in the EC cages singly instead of in groups, for two hours per day for 30 days. Both they and their IC littermates were injected with either methamphetamine or saline during either the light or dark phase of the day. For brain weight, placing the animals singly in EC cages produced small though significant effects, and these were augmented by the dark phase, during which rats are more active, and by methamphetamine. The combination of dark and amphetamine appeared to produce larger effects than either factor alone, especially in the occipital cortex. For AChE, on the other hand, the effects did not appear to be additive, and for ChE only the combination of all three factors, complexity, drug and dark, appeared to be effective, probably because, as has been previously noted, 30 days is usually insufficient for the induction of effects with this enzyme.

Other Stimulants

A variety of other stimulants have also been tested for their ability to augment brain effects. Not surprisingly, d-amphetamine appeared to exert an influence very similar to that of methamphetamine. On the other hand, pentylenetetrazol (metrazol) produced small though significant augmentation of EC weight effects in the dorsal cortex, total cortex and C/S ratio, but unlike the amphetamines and the EC condition itself, did not affect the occipital cortex. Thus pentylenetetrazol was not particularly effective, and the patterning of change differed from the typical EC effect. Totally without detectable effect on brain weight was strychnine, perhaps because, unlike the other stimulants, this drug exerts its major influence below the level of the forebrain.

## Depressants

Since stimulants enhance EC effects, one might expect that depressants would diminish them, especially in view of the behavioral effects of barbiturates described earlier, and such a diminution has in fact been found. Pentobarb sodium, in a dose of 10 but not 5 mgm/kgm, showed a suggestive reduction of brain weights, and a larger series of experiments using phenobarb sodium ( 30 mgm/kgm) showed a highly significant effect. Once again an interaction between drug and environment was evident, since phenobarb brain weight reductions were small in IC animals and only reached significance in the ventral cortex, whereas in complexity subjects they were larger and their regional patterning, which was greatest (-4.7%) in the occipital region and least in the somatosensory, followed that of EC.

The cholinesterase enzymes were also affected, though less markedly. Once again the drug was without effect in the isolate condition. In the EC condition ChE was apparently reduced by the drug concomitantly with brain weight, so that the activity per unit weight showed no effect. For AChE the drug effect was apparently also to reduce the EC increase in total amount of enzyme, but in this case the reduction was proportionately greater for enzyme than for weight, so that the net drug effect was to reduce AChE activity per unit weight. This drug effect of reducing AChE activity per unit weight is initially confusing, since EC per se has a similar effect, but is more readily understandable when viewed in terms of the primary factors (total amount of AChE, and weight of cortex) from which it is derived.

Interestingly, the phenobarb effects could not be detected when the duration of differential exposure was shortened to only 30 minutes per day. This is in contrast to the amphetamine findings, but is understandable in terms of the effects on behavioral activity in the EC cage, since phenobarb only produced significant reductions in activity after the first half hour.

Thus stimulants and depressants which exert their effects throughout the brain, and not just in lower brain and spinal cord regions, appear to be effective in modifying the magnitude of the brain response to environmental complexity. From behavioral observations, it would appear that modifying exploratory behavior may be an important intermediary in eliciting these effects, although it may be, of course, that this is an unrelated concomitant. These findings support the hypothesis that the arousal reaction may be one of the major neurophysiological mechanisms mediating the induction of environmentally induced brain changes (see Chapter 8). Amphetamines tend to increase the arousal reaction and to delay its habituation, while for the barbiturates the opposite is true; hence, from this hypothesis they would be expected to respectively enhance and depress the effects of complexity. Furthermore, since it is postulated that the EC brain effects are produced largely via the eliciting of arousal reactions to the novel stimuli or toys in the environment, it would be anticipated that where stimulant drugs inhibited exploration, an in the older animals, no enhancement of EC effects would occur. As is apparent, the above findings fit wery well with the hypothesized predictions.

## Hormones

One of the mechanisms most frequently suggested to mediate the effects of environment on brain has been hormonal influences. This certainly seems a reasonable suggestion, in view of the clear demonstrations of hormonal changes as a result of infantile stimulation and of the fact that these hormonal effects can apparently in turn modify the brain. Therefore, Rosenzweig et al. (1972) studied the effects of the usual EC-IC rearing procedures on rats that had been hypophysectomized at about 30 days and compared them to normal controls. The ran three experiments using male Fischer and Long-Evans strains differentially reared from about 40 to about 70 days of age. The hypophysectomized animals has significantly lower weights of both cortex (-5%) and the rest of the brain (-10%), lower cortical AChE concentration and greater cortical ChE:AChE than did control rats, but they still showed the usual EC-IC effects on all these parameters.

In another study Diamond (1976) examined the effects of male sex hormones. A group of rats was castrated at birth before the brain was essentially male, and another group was castrated at 30 days of age. Both groups were then placed in enriched or impoverished environments from 60 to 90 days of age, and both groups clearly displayed the environmental effects seen in the intact animal. Clearly, male sex hormones do not play a significant role in environmental effects.

On the other hand, the fact that hormones do not appear to play a major role in eliciting environmental effects does not mean that they may not interact in a variety of ways yet to be determined. Hoover and Diamond (1976) have clearly demonstrated that norethynodrel, a synthetic progesterone-like compound, has such effects. Norethynodrel treatment resulted in reduced cortical depth and less sodium in the visual cortex, but produced no detectable effects on water, chloride, and potassium. Most notable was the fact that this hormone effect was seen only in the enriched animals and appeared to be nonexistent or at least nonsignificant in the isolates.

It seems that while hormones do not appear to play a major role in elicit ing brain effects, it is clear that they are capable of interactions with the environment and can produce significant changes in brain. This fact needs to be taken into consideration in future studies of hormonal influences.

## The Rank Order Effect

Comparisons between different species and strains have clearly demonstrated an interaction between environment and genetic background in determining the magnitude of the brain's response to sensory stimulation. However, it seems that no one found (looked for?) systematic patterns of individual differences in brain responses, until Cummins et al. (1977a) reported what they called the Rank Order Effect. They found that when the forebrains from EC animals were rank ordered according to weight and compared with similarly rank ordered IC forebrains, the EC-IC difference was greatest in the lowest weight subjects.

To account for this they presented a model based on the idea that within both groups there would be some reordering of animals within the rank order profile according to individual differences in responsivity to the environment. This reordering was postulated to occur differentially between groups in such a manner as to result in greater EC-IC differences in animals ending up in the lower portions.

Another hypothesis is, of course, simply that low brain weight animals are more sensitive to variations in sensory stimulation. This might be seen as an example of the law of initial values. It might also fit with a variety of findings which indicate that brain damage induced by a variety of etiological agents usually results in reduced brain size, and that these smaller brains tend to exhibit both behavioral and neural hypersensitivity to variations in sensory stimulation and especially to deprivation (Davenport, 1976; Walsh and Cummins, 1976a; Walsh and Greenough, 1976b; Will et al., 1977).

In a subsequent study of the Rank Order Effect, Cummins (1978) noted a time dependent effect on cerebral length. Examining subjects differentially reared for periods from 18 to 160 days revealed a similar Rank Order Effect in the 18-day and some of the 30-day subjects, but with longer periods a reverse effect tended to occur. That is, in these older animals it was the high cerebral length subjects that displayed an EC-IC difference. To incorporate these findings, Cummins has altered his model to include an age dependent modification of the developmental ceiling which would modify the rank reordering for brain weight. Other explanations which do not involve any rank reordering are also clearly possible, but at the present time it is probably impossible to decide among them.

Unfortunately, Rosenzweig (1978) was unable to confirm the rank order effect for cerebral weight in a number of tests involving both standard EC and IC groups as well as subjects reared in the outdoor, semi-natural environment. He found only a slight trend in the expected direction, apparently largely due to disproportionately low values in the lowest ranking isolates. One possible explanation for this negative finding is that he employed smaller groups (averaging about 11) than did Cummins (1978; Cummins et al., 1977a) (averaging about 17), which might reduce the sensitivity of a rank order effect. In any event, independent confirmation is clearly desirable.

Genetic Influences

With the growing evidence pointing to the dangers of ignoring genetic variables in developmental psychobiology and early experience research (Henderson, 1968), systematic investigation of genetic influences on susceptibility to environmental brain plasticity has recently begun. Most studies have used the rat, and so far all have been confined to rodent species. However, within the rodent family, species, strain, and sex differences and interactions have been sought and found.

*Strain Differences*

It may well turn out that strain differences prove to vary with the species, for while the rat would appear to present a remarkably uniform response across strains, at least in the male, for mice a different picture may be emerging. In the rat, to date, almost no differences (strain x environment interaction) have been found for any of the weight or esterase measures (Rosenzweig et al., 1969). Hybrid males, crossbred between Fisher males and Berkeley $S_1$ and $S_3$ lines, also showed similar patterns (Rosenzweig and Bennett, 1977). However, as we shall see, a more complex picture emerged for female hybrids. An exception is that, unlike other strains, inbred Fisher rats do not appear to develop significant effects by 30 days when EC exposure is limited to two hours per day (Bennett et al., 1973).

Within the mouse, however, the only other species for which strain comparisons are available, different patterns appear. LaTorre (1968) compared A/Crgl and C57BL/Crgl mice and found a difference in regional response magnitude. For the A strain, as with the rat, the occipital region showed the greatest response in both weight and AChE, whereas for the C57 the effects were greatest in ventral cortex and small and nonsignificant in the occipital.

Henderson (1970a; 1973) has used the diallel cross method to perform a quantitative assay of genetic influences on whole brain weight of mice. Using six inbred strains as well as their F1 and F2 crosses, he found marked evidence of inbreeding depression (directional dominance), since only the hybrid animals exhibited significant EC-IC brain weight modulation and hybrids had greater brain weights than inbred animals in the same environments. If confirmed, this study has considerable implications for work in this field. Firstly, it suggest that future studies of enrichment, at least for the mouse, would be best advised not to use the inbred animals that have previously been the norm.

Furthermore, when a major portion of genetic variance is determined by dominant genes, this is strong evidence of selective pressure and relevance for survival (Roberts, 1967); and when the dominance is largely unidirectional, then the characteristics of hybrids as compared to inbreds indicate the optimal direction of change. Thus it would seem that the ability to respond to different environments with corresponding changes in brain size confers definite survival value. It should be noted that these findings suggest not only that there is an advantage in having a larger brain in a complex environment, but also that it may be adaptive to have a smaller one in simpler times. In any case, these results clearly show that inferences concerning the adaptiveness of a characteristic may be quite misleading if either the laboratory population or its environment differ significantly from natural populations or environments (Henderson, 1973).

## *Sex Differences*

Almost invariably experimenters have worked with male subjects, presumably in an attempt to avoid confounding effects of cyclic hormonal variations. This would seem a wise precaution, since estrogenic and progestogenic substances are known to exert excitatory and depressive effects respectively, and to be capable of modifying CNS maturation (Timiras, 1972; Wooley et al., 1960).

However, only recently has evidence of sex differences in brain response to environmental enrichment and deprivation become available. In the rat hybrids discussed above, males appeared to display greater EC-IC effects than females after the confounding influence of body weight had been removed (Rosenzweig and Bennett, 1977), The regional patterning of cortical effects also showed differences, in that in the occipital region males displayed larger effects whereas the reverse was true in the somesthetic (Diamond, 1976; Rosenzweig and Bennett, 1977).

Furthermore, it seems that enrichment and isolation exert somewhat different effects on males and females. Thus in the case of the C/S ratio, the enrichment and impoverishment effects, compared to the standard colony baseline, were similar in males. In females, however, the enrichment effect was considerably larger than the impoverishment. For ChE:AChE, on the other hand, the reverse was true, since females now displayed a significantly larger impoverishment than enrichment effect. In general females appear to be less susceptible to environmental effects, especially impoverishment.

Pregnancy and breeding are now known to induce significant cerebral effects. Diamond et al. (1971) reared male and female rats differentially for 56 days, and during an interim period of 5 days paired either EC males and females or IC males and females. Another group of EC and IC males and females were similarly reared, except that they were paired with partners of the same sex during the 5 day period.

Cortical depth measurements revealed significant EC-IC differences for both males and nonpregnant females, although they displayed a differential regional response specificity. Thus males revealed a greater response in the occipital region, especially medially, whereas nonpregnant females exhibited significant effects in the somesthetic region, where changes in the males were nonsignificant. However, in the postpartum females no significant cortical depth effects of differential rearing could be detected. The effect of the procedure of mating, pregnancy and parturition was to increase cortical depth, especially in the frontal cortex but also, though to a lesser extent, in the occipital and somatosensory regions. This effect was greater in IC than in EC animals, thus largely eradicating the effects of differential rearing. The neonate pups of these matings were also examined, and it was found that in both the male and female offspring of the EC adults body weight was significantly greater, but no effect on the cortical depth was observed.

*Sex X Strain Interaction*

As mentioned in the discussion of strain differences, an attempt to replicate Henderson's (1970a) hybrid study using male rats was unsuccessful, since two kinds of hybrids of $S_1$ X Fisher and $S_3$ X Fisher did not differ from their parent strains (Rosenzweig et al., 1973; Rosenzweig and Bennett, 1977). However, hybrid females were slower to develop enrichment effects (EC-SC differences) and showed only minimal isolation effects at both 30 and 80 days. Thus, to some extent, the female hybrid rat was less susceptible to environment and was partially protected against the effects of isolation. This brings to mind the reports of Sackett (1972) and Cummins et al. (1974a, 1977b; Worsley et al., 1973). Sackett found that in monkeys, males were both more susceptible to the deleterious effects of isolation, as manifested behaviorally by their decreased tendency to explore and their increased sterotypies, and also more likely to improve in a subsequent social therapeutic environment. Simmilarly, Cummins et al. (1976b) found that prior social environment was a much more potent determinant of subsequent cheeping behavior in the male than in the female chick when they were placed in a novel environment. In general, then, it appears that at least for the noninbred strains the female is less susceptible to both behavioral and the cerebral effects of social environment; in particular she seems to be to some extent protected against the deleterious effects of isolation.

*Species Differences*

As yet only limited studies of the mouse (LaTorre, 1968), gerbil (Rosenzweig and Bennett, 1969), and feral deermouse (Rosenzweig and Bennett, 1977) are available, so that comparisons between species are necessarily limited. It would seem that nearly all parameters change in a qualitatively similar manner as in the rat, but that magnitude and regional specificity vary.

Cortical weight showed positive EC-IC differences in all three species, but whereas for the rat this effect was most marked in the occipital region, in the others this was not necessarily so. For the C57 mouse the effect was least marked occipitally and maximal in the ventral cortex, whereas for the gerbil changes appeared to be relatively uniform throughout. This absence of a marked occipital EC-IC weight difference in the gerbil seems to be due to the fact that the IC-SC difference is positive, rather than negative as in the rat. If this finding is confirmed it will greatly reduce the generality of the monotonicity hypothesis of increasing brain change with increasing levels of environmental complexity (EC > SC > IC) that has been implicit in this field since its inception (Krech et al., 1960).

AChE cortical changes were similar across species. However, in the subcortex, (whereas for the rat AChE/weight exhibits a positive EC-IC effect) the opposite would seem to hold for the gerbil. For ChE, however, the mouse would appear to show the greatest difference from the rat, since not only was ChE activity not increased throughout the cortex in EC, but there was in fact a significant reduction in the occipital region.

The interpretation of all these genetic effects is difficult at this stage; whether they represent direct genetic influences on the patterns and degree of brain plasiticity, or whether they are secondary to such things as differing behavioral responses to environment, is unknown. While the available evidence indicates that the structural and chemical responses of the brain to environmental complexity and impoverishment are not limited to any one species, sex, or strain, the precise nature of the genetic-environment interaction remains almost entirely unknown. The application of precise quantitative genetic techniques, such as the diallel cross, to a wide range of cerebral parameters would seem a necessary step, and will perhaps yield precise information, as it has already done for the behavioral effects of enrichment.

There is an obvious need for extension to nonrodent species, preferably higher mamals, since it may be that one will find increased effects as one ascends the phylogenetic scale, and examines animals capable of more complex behaviors (Rosenzweig, 1966). One of our ultimate aims must be to study primates, in the hope of extrapolating more directly to man.

## Large Brains or Small: Pros And Cons

As one ascends the phylogenetic scale, one clearly finds that brain size, neural interconnections, and behavioral plasticity tend to increase, and it is intuitively obvious that larger brain size may well confer adaptive advantages. There are also obvious intuitive advantages which might be expected to follow from the cerebral enhancement in size which a complex environment elicits. Greater behavioral plasticity, adaptiveness, and memory and learning skills are obvious possible concomitants. One particularly interesting question is to what extent are the brain changes specifically and optimally adaptational for that environment?

While the advantages of a larger brain seem obvious, it is not so obvious that there may also be advantages to a smaller brain. One advantage is energy expenditure, since the cells of the CNS continuously operate at a high metabolic rate and thus account for a disproportionate amount of the body's total energy use; for example, the human brain represents only 2% of body weight yet requires 20% of the energy expenditure. Therefore, when an animal is not faced with a demanding environment, there may be a selective advantage in reducing the activity and size of the nervous system, in the direction of minimal but functionally adequate values. This would be consistent with survival, as long as the animal retained the ability to enlarge the brain should the environment suddenly become more demanding. It is therefore of interest to note that brain changes appear more marked and may possibly occur with greater rapidity after a period of stimulus deprivation (Cummins et al., 1973). Thus the ability to modify brain size in either direction may, by virtue of compromising with metabolic costs, confer selective advantage upon a population (Rosenzweig, 1978).

CHAPTER 6

# Effects of Social Environment on Cerebral Morphology, Chemistry and Pharmacology

Francis V. De Feudis, Ph.D.
Institut Henri Beaufour
Les Ulis, France

## INTRODUCTION

This chapter examines the effects of changes in the level of social environmental stimulation (experience) on cerebral morphology, chemistry and pharmacology in mammals. Some of these effects of environment have been monitored in humans as well as in experimental animals.

In rodents, certain environmental restrictions (e.g., isolation-rearing) produce effects on cerebral morphology and chemistry, and on the brain's responses to psycho-active agents, as well as altering behavior. In mice, such effects have been termed the "isolation syndrome"(e.g., Valzelli, 1973a). Male mice become aggressive after being exposed to prolonged social isolation (i.e., individual housing for 4 - 12 weeks; see Allee, 1942; Scott and Fredericson, 1951; Yen et al., 1959; Janssen et al., 1960). On the other hand, enrichment, or colonial-style living among rodents, which increases environmental stimulation, causes different types of symptomatology. Several anatomical and biochemical parameters of the CNS are also modified by such treatment (Bennett and Rosenzweig, 1971), and these changes have been correlated with an enhancement of learning ability and problem-solving behaviors (Forgays and Forgays, 1952; Hymovich, 1952; Krech et al., 1962a, 1962b; Rosenzweig, 1964). Thus an impoverished environment, which reduces an animal's level of social interaction, generally acts oppositely on brain and behavior than do conditions of scoial enrichment. Aggressive tendencies, however, do exist in colonies of rodents, especially when a social hierarchy is being formed (Christian, 1955; Louch and Higginbotham, 1967).

The relevancy of these studies in explaining the etiologies of behavioral modifications in humans cannot be over-emphasized. Some degree of social isolation (i.e., some decrease in social interaction) could be involved in all behavioral changes which occur in animals, including man. Altered behaviors (e.g., depressive states) which are caused by a lack of physical and emotional contact with con-specifics may be due to a functional deafferentation of the somatosensory system (Prescott, 1971; Riesen, 1961).

Much study will be required before the interactions among environment, brain and behavior can be adequately discussed. The effects of biological rhythms, species, age and sex differences, and hormonal changes on these parameters should be seriously considered. Although many of these elements have not yet been systematically studied, an effort will be made here to show some relationships which exist between environmentally induced cerebral morphological and biochemical changes and behavior.

## EFFECTS OF ENVIRONMENT ON CEREBRAL MORPHOLOGY

### Changes In Brain Weights

Krech and his colleagues (1962), Bennett and Rosenzweig (1971) have shown that rodents raised in enriched or social environments possess heavier and thicker cerebral cortices than isolates (see Table 1 and Chapter 5). Also the forebrains of aggregated mice weighed slightly less than those of their isolated littermates (DeFeudis, 1974a).

Table 1
Effects of Differential Experience on Occipital Cortex of $S_1$ Rats Kept in "Enriched" or "Impoverished" Conditions from 25 to 105 Days of Age.

| Parameter | % Difference (EC > IC) | p-value | Number (EC > IC) |
|---|---|---|---|
| Weight | 6.4 | < 0.001 | 133/175 |
| Total protein | 7.8* | < 0.001 | 25/32 |
| Thickness | 6.3 | < 0.001 | 45/52 |
| Total AChE | 2.2 | < 0.01 | 102/171 |
| Total ChE | 10.2 | < 0.001 | 118.5/132 |
| Total hexokinase | 6.9+ | < 0.01 | 17/21 |
| DNA/mg | -6.1 | < 0.001 | 4/23 |
| RNA/mg | -0.7 | N.S. | 10/23 |
| RNA/DNA | 5.9 | < 0.01 | 19/23 |
| No. neurones | -3.1 | N.S. | 7/17 |
| No. glia | 14.0 | < 0.01 | 12/17 |
| Perikaryon cross section | 13.4 | < 0.001 | 11.5/13 |

EC = enriched animals; IC = impoverished animals
* Weight difference 7.0% in these experiments
\+ Weight difference 5.5% in these experiments
(From Bennett and Rosenzweig, 1971)

### A Change In Brain Water

Another index of morphological change is simply the water content of the brain. If cerebral water content is altered, then it seems reasonable to believe that the volume of the intra- and/or extra-cellular compartment(s) has changed. This type of alteration might reflect changes in the sizes (or relative sizes) of glial vs. neuronal compartments, or a change in the size of the functional extra-cellular space. Only a few studies have been conducted on the possible influence of environment on this parameter. In recent work, male, weanling (21-22-day-old) C57 Black mice were housed either singly ("isolated") or in groups of 25 ("aggregated") for 3-6 weeks. After their sacrifice, the whole left side of the brain (rostral to the inferior colliculi, excluding cerebellum) and the right cerebral cortex were excised rapidly and prepared for water determination (DeFeudis and Elliott, 1967). The brains and cerebral hemispheres of isolated mice contained less water than those of their "aggregated" littermates (DeFeudis, 1972a). Previously, Bennett et al. (1969) found no difference between dry:wet weight ratios of the brains of rats subjected to "enriched" vs "impoverished" conditions. The discrepancy was probably due to the differences in species or housing conditions, or in the methods used for dry weight determinations. The lower % brain water content of "isolated" mice (DeFeudis, 1972a) may be related to the decreases in glial elements which have been shown to occur in the brains of rodents subjected to impoverished housing conditions (Altman and Das, 1964; Bennett and Rosenzweig, 1971). Glial structures, which may represent a portion of the functional extra-cellular space of brain, might be altered in differentially-housed mice to an extent which can account for the difference in % water of their brains.

### Changes In Cerebral Microscopic Anatomy

The changes in brain weight, protein content, enzyme activities and nucleic acid levels reported for animals subjected to differential housing do indeed indicate that the microscopic anatomy of the brain changes with respect to the environment.

After changes in brain weight had been reported by Krech and his co-workers (1962a, 1962b), it seemed pertinent to study these changes in terms of specific cellular populations. In this regard, Altman and Das (1964) determined the incorporation of $^3$H-thymidine in to the brains of 4-month-old rats (raised since weaning under "enriched" vs. "isolated" environments) one week after its administration. "Enriched" rats possessed 60% more labeled "glia-like" elements in their neocortices than "isolated" rats. These labeled cells appeared to be primarily undifferentiated elements migrating from the sub-ependymal germinal zone of the lateral ventricle to the cortical grey matter (Altman, 1967). Regardless of the type of elements involved, these results revealed that environmental changes could significantly alter cerebral morphology. Other studies (see Rosenzweig, 1967) revealed that the numbers of neurones per microscopic field, within specific regions of the occipital cortex, were slightly decreased in "enriched" rats, suggesting a dilution effect within an expanding cerebral cortex. Both glial number and the glia/neurone ratio were increased. Cross-sectional areas of the perikarya and nuclei of neurones were also shown to be significantly increased in the cortices of "enriched" rats (Diamond et al., 1966; Diamond, 1967; see also Diamond et al., 1975). The histology of rat hippocampus was also altered by environmental complexity (Walsh et al., 1969).

## EFFECTS OF ENVIRONMENT ON CEREBRAL CHEMISTRY

### Nucleic Acids

#### *Deoxyribonucleic Acid (DNA)*

Many studies regarding the effects of learning, training and environment on cerebral RNA content and turnover have revealed that changes in behavior are often reflected by alterations in brain RNA and protein metabolism (e.g., Hydén, 1967; Glassman, 1969; Rose, 1969; DeFeudis and DeFeudis, 1977). However, relatively few studies have been conducted on the effects which the environment may produce on cerebral DNA.

Altman and Das (1964) injected the DNA precursor, $^3$H-thymidine, intraperitoneally into rats which had been raised under "enriched" vs. "impoverished" conditions for about 3.5 months. Upon sacrifice of the animals a week later, the size of the neocortex of "enriched" animals was increased and more newly formed glia-like elements were labeled. Essman (1971b) found that the diploid cell number (and hence the DNA content) was significantly reduced, especially in diencephalon and cerebellum, in mice that had been "isolated" between the ages of 21 and 49 days. More recently, Welch et al. (1974a) showed that after one year of differential housing, "grouped" rats had less telencephalic DNA than "isolated" rats. They suggested that, under some conditions, environment may affect the age-associated loss of cerebral neurones and/or the metabolism of satellite DNA.

*Rubonucleic Acid (RNA*

Although early reports indicated that cerebral RNA was not altered by rearing rats in "enriched" environments (Bryan, 1965; Rosenzweig, 1966), more recent studies have revealed that cerebral RNA does undergo change in rats exposed to "enriched" environments (Ferchmin et al., 1970; Bennett, 1976) and in differentially housed mice (e.g., Essman, 1971a, 1972).

The most dramatic changes have been reported by Essman (1971a), who found that the isolation of male mice produced an increase in glial RNA which occurred simultaneously with a decrease in neuronal RNA. Essman (1971a, 1971b) further showed that environmental stimulation markedly increased "% glial RNA concentration" in the brains of "isolated" mice, while not affecting it in "grouped" mice. It was suggested that the changes in the glial cell number caused by isolation of mice could represent either glial agenesis, which is age- or development-dependent, or delayed glial proliferation.

Regarding studies with rats, Ferchmin et al (1970) reported an increase in cerebral RNA content after four days of environmental enrichment. These results were confirmed, and, in addition, a significantly higher RNA:DNA ratio was shown to exist in the cerebral cortices of "enriched" rats, as compared with those of their "impoverished" littermates (Bennett, 1976). Grouped rats also appear to have had more telencephalic RNA than "isolated" rats (Welch et al., 1974a). Most recently, Grouse et al. (1978) have shown that changes in sequence diversity of brain RNA can be caused by exposing rats to environmental complexity.

These studies on nucleic acids, together with those on proteins (see below), have revealed that considerable biochemical plasticity exists in the brains of young rodents. This could be a factor involved in environmental adaptation.

Proteins

Cerebral proteins might undergo changes in response to environmental modification. Enzymes involved in neurotransmitter synthesis and degradation are of prime importance. When the level of a neurotransmitter substance, or energy metabolite, of the brain is altered by environmental exposure, it is expected that concomitant changes occur in the associated enzyme systems, but such alterations have been difficult to demonstrate.

The most extensive analyses of changes in cerebral enzymes which are caused by environmental modification have been performed on cerebral cholinesterases (Bennett and Rosenzweig, 1971; Central Cholinergic Systems Section, below). It has been maintained that rodents raised in "enriched" environments have more ChE in subcortical structures and less AChE in the cerebral cortex (Rosenzweig et al., 1962, 1969; Bennett and Rosenzweig, 1971; Fig. 1). In the three species examined (rat, mouse, gerbil), AChE/unit weight of cerebral cortex was lower for "enriched

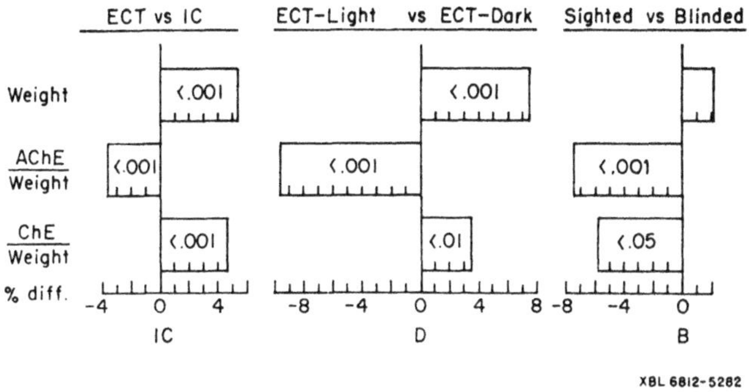


Fig. 1. Effects of 3 sets of treatments on weight and enzymatic activity of occipital cortex of $S_1$ strain rats. All treatments were maintained from 25 to 105 days of age. ECT-IC results are based on 98 littermate pairs. Results of comparing ECT in normal light cycle with ECT kept in complete darkness are based on 31 littermate pairs. The Sighted-Blinded comparisons for weight and AChE are based on 20 litters, each litter furnishing 2 sighted rats (S-ECT and S-IC) and 2 blinded rats (B-ECT and B-IC); the Sighted values combine S-ECT and S-IC, and the Blinded values combine B-ECT and B-IC. The ChE Sighted-Blinded value is based on only 9 litters of 4 rats; that is, 18 sighted and 18 blinded rats. Percentage differences between the first and second treatment of each set are referred to the second member (IC, Dark, Blinded) as baseline. The statistical significance of each difference is shown in the appropriate bar; only the weight difference between Sighted and Blinded was not significant; ECT= environmental complexity plus maze training. (From Rosenzweig et al., 1969).

than for "impoverished" animals; this effect was most pronounced in the occipital cortex (Rosenzweig and Bennett, 1969; Table 2). Also, for all three species, "enriched" animals exceeded "impoverished" animals in ChE/AChE ratio of total cortex (Rosenzweig and Bennett, 1969). But, more recent studies have indicated that cerebral AChE activity can be altered by other means. Treatment with stimulant drugs (e.g., methamphetamine, d-amphetamine) can enhance the effects of "enriched" experience on the cerebral activities of AChE and ChE in male rats (Bennett et al., 1973). Interestingly, although cholinesterases appeared to change in relation to environmental modification, choline acetylase (measured with the method of McCaman and Hunt, 1965) was not altered in male Swiss mice subjected to prolonged periods of individual housing (Consolo and Valzelli, 1970). Other cerebral enzymes have not received much study to date. A slight increase in total hexokinase activity of occipital cortex of rats may have been caused by "enriched" experience (see Rosenzweig,1967). Other reports (e.g., Rosenzweig, 1967) indicated that a slight increase in total cortical protein content occurred in rats which had undergone "enriched" experience. More recently, it has been shown that cerebral nerve ending fractions of "isolated" male mice have less protein than those of their "aggregated" littermates (DeFeudis, 1972e; DeFeudis et al., 1976a). Thus, environmental impoverishment might cause a decrease in the number, or size, of cerebral nerve endings (see also DeFeudis, 1977).

With regard to protein synthesis, Altman and Das (1964) found no difference in the cerebral uptake of injected $^3$H-leucine between "enriched" and "impoverished" rats. However, more recent subcellular studies with $^3$H-leucine have revealed that proteins of brain nuclei and synaptic endings appear to be selectively affected by environmental complexity (e.g., Levitan et al., 1972a,b,c). Synaptosomal proteins of hippocampus incorporated $^3$H-leucine at a higher rate in "enriched" than in "deprived" rats, and the opposite effect occurred in the cortex. In another study, increased incorporation

Table 2. Percentage EC-IC Differences in Brain Weights and Brain Enzymes of Rats, Mice, and Gerbils Differentially Reared for 30 days.

| | $S_1$ Rat | | Mouse: A strain | | Mouse: C57 strain | | Gerbil | |
|---|---|---|---|---|---|---|---|---|
| | (N=29 per group) | | (N=12 per group) | | (N=12 per group) | | (N=23 per group) | |
| | % diff. | P | % diff. | P | % diff. | P | %diff. | P |
| **Weight** | | | | | | | | |
| Occipital cortex | 10.3 | < .001 | 19.6 | < .01 | 5.8 | N.S. | 4.0 | <.10 |
| Total cortex | 5.8 | < .001 | 10.7 | < .001 | 8.2 | <.001 | 4.4 | <.001 |
| Subcortex | -0.1 | N.S. | 2.7 | N.S. | 2.0 | N.S. | 1.7 | <.10 |
| Cortex/Rest | 5.9 | < .001 | 7.6 | < .01 | 6.0 | <.01 | 2.6 | <.05 |
| **AChE/weight** | | | | | | | | |
| Occipital cortex | -4.8 | < .001 | -7.8 | < .01 | -5.0 | <.10 | -5.2 | <.001 |
| Total cortex | -3.2 | < .01 | -5.1 | N.S. | -1.5 | N.S. | -3.2 | <.05 |
| Subcortex | 2.5 | < .05 | 0.9 | N.S. | 0.6 | N.S. | -1.4 | <.10 |
| Cortex/Rest | -5.5 | < .001 | -5.6 | N.S. | -2.1 | N.S. | -1.8 | N.S. |
| **ChE/weight** | | | | | | | | |
| Occipital cortex | 1.8 | −.10 | -14.5 | < .01 | -1.2 | N.S. | 2.8 | <.10 |
| Total cortex | 1.2 | N.S. | -0.8 | N.S. | 2.7 | N.S. | -0.7 | N.S. |
| Subcortex | 1.2 | < .10 | -0.3 | N.S. | -3.2 | N.S. | -1.0 | N.S. |
| Cortex/Rest | 0.0 | N.S. | -0.4 | N.S. | 5.3 | N.S. | 0.5 | N.S. |
| **ChE/AChE** | | | | | | | | |
| Occipital cortex | 6.9 | < .001 | -7.0 | < .10 | 3.0 | N.S. | 8.8 | <.001 |
| Total cortex | 4.7 | < .001 | 4.6 | N.S. | 4.1 | N.S. | 2.5 | N.S. |
| Subcortex | -1.2 | N.S. | -1.1 | N.S. | -3.8 | N.S. | 0.4 | N.S. |
| Cortex/Rest | 6.0 | < .001 | 6.2 | N.S. | 7.3 | N.S. | 2.4 | N.S. |

(From Rosenzweig and Bennett, 1969)

of $^{14}C$-leucine into the hippocampi of rats occurred with four days of training (in the use of the non-preferred paw in retrieving food). This change occurred in all subcellular fractions of hippocampus except perikaryal mitochondria (Levitan et al., 1972c). Hence, an increased metabolic demand appears to exist in the hippocampus when rats are trained for four days at this task. For further discussion of the effects of environment on cerebral protein synthesis, see DeFeudis and DeFeudis (1977).

## Lipids

To date, no systematic studies have been performed on the possible effects which environmental modification might exert on cerebral lipids. However, it seems likely that such changes do occur, especially since differential housing of mice causes marked effects on the incorporation of carbon atoms of glucose and other energy substrates into cerebral structures (DeFeudis 1971, 1972b, 1973, 1974c). Recent studies which have revealed that the "binding" of possible transmitter substances and psycho-active agents is affected by differential housing (DeFeudis 1972d, 1972f; Hendley et al., 1973; DeFeudis et al., 1976b) also lead one to suspect that complex lipids of synaptic membranes are altered by environmental modification.

## Extra- and Intra-Cellular Fluids

Results which revealed that the "% water" content of brain is modified by differential housing (DeFeudis, 1972a; see Section above) lead one to suspect that such treatment might produce changes in cerebral extra- and intra-cellular spaces, and provide some support for reported changes in glial cell number (Essman, 1971a). Other studies which revealed that a greater amount of $Li^+$ was retained by cerebral subcellular fractions of "isolated" mice, than by those of their "aggregated" littermates (DeFeudis, 1972f), also indicated that cerebral fluid composition might be altered by environmental modification (see also Welch et al., 1974a).

## Carbohydrate Metabolism

### *Introduction*

Early studies indicated that "depression" in humans was accompanied by a decrease in glucose tolerance (Kooy, 1919; Mann, 1925), and today it seems well established that abnormalities in carbohydrate (glucose) metabolism are associated with many emotional and mental distrubances of man (see references in Henneman et al., 1954; Shopsin et al., 1972; Sacks, 1973). Since most mammals have a cerebral respiratory quotient (R.Q.) that is close to unity (indicating that under resting conditions cerebral metabolic energy is derived rather exclusively from carbohydrate), it seems that the causes of behavioral changes which occur in these animals (including man) can be studied by subjecting them to conditions of social isolation and then monitoring the changes produced in glucose carbon atom incorporation into their brains or other organs. Differentially housed mice can be used for such studies.

### *Studies on Carbohydrate Utilization in Differentially Housed Mice*

Individually housed mice have lower resting metabolic rates, weigh more, and exhibit lower locomotor activity than their grouped counterparts when both sets of animals are in their home cages (Welch, 1965; Welch and Welch, 1969a). Carbohydrate metabolism in mice can be altered by changing their living conditions.

From several recent studies (DeFeudis, 1971, 1972b, 1973; DeFeudis and Paolino, 1972), it has become evident that individual housing of male mice causes a marked decrease in the cerebral incorporation of carbon atoms of glucose (or other energy substrates) (Table 3).

In these studies, male, weanling, C57 Black or Swiss-Webster mice were used. "Isolated" mice were housed singly in small cages in a quiet room, and "aggregated" mice were kept in large cages in groups of 20-30 in a noisy room with other animals. No "enrichment" was provided to the "aggregated" mice, and all mice were housed differentially for 3.5-11 weeks. Decreases in the incorporation of radioactive carbon atoms of subcutaneously administered $U^{14}C$-D-glucose or $1^{14}C$- pyruvate occurred not only in the brain at 30 min. post-injection, but also in the liver (with glucose or pyruvate) and in other tissues (tested only with glucose).

In studies involving radioactive glucose and fructose, mice were fasted for 12-18 hours prior to sacrifice; with radioactive pyruvate the differences in cerebral carbon atom incorporation, between "isolated" and "aggregated" mice, were similar, regardless of whether or not the animals were fasted. Differences in cerebral glucose carbon atom incorporation were in the order of 50-60% ("aggregated" > "isolated"). Carbon atoms of injected $^{14}C$-inulin (which is not involved in energy metabolism) were incorporated to a similar extent into the brains of "isolated" and "aggregated" mice (Table 3). The difference in glucose carbon atom incorporation between "isolated" and aggregated" mice occurred rather uniformly among several brain regions and exhibited a diurnal rhythmicity (DeFeudis and Black, 1972). The difference was smallest at 1800-2000 hr. which was also the period during which both groups of mice exhibited maximal motor activity.

Since carbon atoms of $1\text{-}^{14}C$-D-mannose enter mouse brain at least as readily as those of $U^{14}C$-D-glucose (DeFeudis, 1972c), and since perfused rat brain preparations metabolize mannose and glucose in similar fashion (Sloviter and Kamimoto, 1970), this monosaccharide was used in further studies. The amount of radioactivity of mannose incorporated into the brain was about 30-35% greater than that of glucose for both "isolated" and "aggregated" mice (DeFeudis, 1972b). In these studies, the mice were killed 30 minutes after injections of radioactive substances. Further information will be available after studies on the time courses of these effects.

These studies have supported the liaison between decreased glucose tolerance and behavioral changes, since the behavioral changes produced by subjecting animals to social isolation appear to be governed by changes in the rate of glucose utilization.

What do these studies mean, in more specific terms? Although it is not yet clear, the decreased incorporation of carbon atoms into the brains of "isolated" mice may be due to: 1) decreased energy metabolism; 2) decreased synthesis of cerebral and liver glycogen (see Welch, 1965); 3) increased peripheral tissue metabolism; or 4) alterations in serum levels of hormones that regulate carbohydrate metabolism (e.g., insulin, 11-oxysteriods, epinephrine, glucagon, ACTH, thyroxine). Since differences between "isolated" and "aggregated" mice varied diurnally, and since carbohydrate metabolism also undergoes diurnal variation (Sollberger, 1960, 1965), it seems apparent that further studies on hormonal rhythms will greatly amplify these findings. Also, there seems to be little doubt that some postulated neurotransmitter substances that are synthesized _de novo_ from glucose in the brain (e.g., acetylcholine, glutamate, GABA) are affected by environmental changes. Hence, upon further analysis of the effects that differential housing causes on cerebral glucose metabolism (in terms of the syntheses of postulated neurotransmitter substances), more direct information relating environmental changes to the neurochemistry of behavior should become available.

Table 3. Incorporation of Radioactive Carbon Atoms of Glucose, Fructose, Pyruvate and Inulin into the Brains of Differentially Housed, Male C57 Black Mice.

| Days of "isolation" or "aggregation" | Substance injected | Radioactivity in Brain dpm/g ($X10^{-3}$) Homogenate | Extract |
|---|---|---|---|
| | "Isolated" mice | | |
| 27 | $U^{14}C$-D-glucose | 304 ± 7 (5)* | 306 ± 6(5)* |
| 27 | $U^{14}C$-D-fructose | 273 ± 7.5(5)* | 273 ± 5.5(5)* |
| 35 | $1^{14}C$-pyruvate | 30 ± 2(8)* | 17 ± 0.6(7)* |
| 27-35 | $U^{14}C$-inulin | 11 ± 0.2(4) | 12 ± 1.8(4) |
| | "Aggregated" mice | | |
| 27 | $U^{14}C$-D-glucose | 474 ± 17(6) | 468 ± 16(6) |
| 27 | $U^{14}C$-D-fructose | 394 ± 11(6) | 395 ± 12(6) |
| 35 | $1^{14}C$-pyruvate | 49 ± 1.7(6) | 35 ± 0.6(6) |
| 27-35 | $U^{14}C$-inulin | 12 ± 1.7(4) | 13 ± 2.2(4) |

Mice were fasted 12-18 hours before experiments and were sacrificed, either by immersion in liquid $N_2$ or by decapitation, 30 minutes after subcutaneous injections of "tracer" amounts of radioactive substances. Values are presented for homogenates of brain and for 80% ethanolic extracts of these same homogenates. Means ± standard errors; numbers of mice in parentheses, * indicates $p < 0.001$ when these values are compared with corresponding values for "aggregated" mice (Student's $t$ - test; 2 - tailed).
(From DeFeudis, 1971; DeFeudis and Paolino, 1972)

*Interactions of environment, psychoactive agents and cerebral carbohydrate metabolism*

Administration of $Li^+$ (200-400 mg/kg, i.p.), 3 hours before subcutaneous injection of a "tracer" dose of $1^{14}C$-pyruvate, caused an increase in carbon atom incorporation which was *selective* to the brains and livers of "isolated", male C57 Black mice. With a 400 mg/kg dose of $Li^+$, carbon atom incorporation into tissues of "isolated" mice was indistinguishable from that of their "aggregated" littermates (Table 4). This "normalization" effect of $Li^+$ is of interest, especially in view of the hyperarousability of "isolated" mice and the clinical effectiveness of $Li^+$ as an anti-manic agent.

Injected d-amphetamine decreased *selectively* the incorporation of carbon atoms of $U^{14}C$-D-glucose and $1^{14}C$-D-mannose into the brains of "isolated" mice (DeFeudis et al., 1972; DeFeudis, 1973). Hence, this psychomotor stimulant acted oppositely to the "anti-manic" agent, $Li^+$, whereas both agents acted *selectively* on carbohydrate metabolism of the "isolated" mice. These results agreed with previous studies which had revealed that d-amphetamine was more toxic to long-term "isolated" mice than to their "grouped" counterparts. Welch (1965) suggested that this difference was probably caused by the greater excitability of the "isolated" mice, and that perhaps it reflected their greater release of endogenous catecholamines in response to the excitement caused by handling and drug injection.

Although the above-mentioned studies indicated that d-amphetamine exerted a selective action on glucose metabolism in "isolated" mice (an effect which could explain its enhanced toxicity in these animals), it should be understood that this effect was due to "prolonged" periods (e.g., 3.5-12 weeks) of differential housing. Chance (1946) showed that "short-term" exposure of mice to "aggregated" conditions potentiated the excitatory effects and lethality of injected d-amphetamine (see also Weaver and Kerley, 1962; Moore et al., 1965; Fuller and Hines, 1967). This effect was also correlated with carbohydrate metabolism (e.g., Moore et al., 1965; Bewsher et al., 1966; Clark et al., 1967). Moore et al. (1965) showed also that d-amphetamine exerted little effect on blood glucose or on brain, skeletal muscle and liver glycogen levels in short-term "isolated" mice, while markedly reducing the glycogen contents of these tissues in short-term "aggregated" mice.

An interesting finding emerged from a study of the effects of d-amphetamine on the incorporation of carbon atoms of $1^{14}C$-D-mannose into the brains and sera of differentially housed male, Swiss-Webster mice (DeFeudis, 1973). The effects of prolonged changes in environment (e.g., lower cerebral incorporation of carbon atoms in "isolated" mice) were altered markedly by short-term (1.5 hr.) changes in environment. The action of injected d-amphetamine on carbon atom entry into brain and serum was also modified by "short-term" changes in environment. This study revealed that: 1) significant changes in cerebral metabolism can be caused by "short-term" changes in environment; 2) "short-term" environmental changes caused similar effects on brain metabolism in both "isolated" and "aggregated" mice; and 3) significant decreases in serum radioactivity caused by d-amphetamine injection occur only in mice that are kept "aggregated" during experiments, regardless of their previous long-term housing conditions. This last effect is due perhaps to an increased utilization of energy substrate by peripheral tissues. The most dramatic changes in serum radioactivity occurred in "isolated" mice which were kept "aggregated" during the experiments, and these coincided with intense fighting behavior. In short, these results provided good evidence that some changes produced by prolonged exposure of mice to differential environments can be reversed by "short-term" exposure to new environments.

Table 4. Effects of LiCl Injections on the Incorporation of Carbon Atoms of $1^{14}C$-Pyruvate into the Brains and Livers of Differentially Housed Mice.

| Injection | Radioactivity in tissue extract*; dpm/g, fresh weight (X $10^{-3}$) "Isolated mice" | "Aggregated mice" |
|---|---|---|
| | Brain | |
| Control; NaCl (550 mg/kg) | 18.1 ± 0.96 (4) | 26.7 ± 3.23 (6)** |
| LiCl (200 mg/kg) | 23.4 ± 2.61 (4) | 28.8 ± 3.43 (6) |
| LiCl (400 mg/kg) | 29.2 ± 3.52 (4)+ | 27.9 ± 2.48 (5) |
| | Liver | |
| Control; NaCl (550 mg/kg) | 87.7 ± 6.55 (4) | 123.4 ± 13.47 (6) |
| LiCl (200 mg/kg) | 73.0 ± 7.54 (4) | 149.5 ± 21.87 (6)++ |
| LiCl (400 mg/kg) | 105.5 ± 5.15 (4)+ | 128.2 ± 15.62 (5) |

Weanling, male C57 Black mice were housed either singly, or in groups of 25, for 10-11 weeks. Three hours before subcutaneous injection of a "tracer" dose (0.1μCi/g) of $1^{14}C$-pyruvate, animals were injected intraperitoneally with either NaCl or LiCl in the doses indicated. All mice were sacrificed 30 minutes after receiving labeled pyruvate. * - Homogenates of brain and liver were extracted with 80% ethanol solution (v/v). +, and ** indicate $p < 0.05$ and $p < 0.025$ respectively, with respect to "isolated" mice, ++, indicates $p < 0.05$, with respect to the "isolated" group which received the same dose of LiCl (Student's t-test; one-tailed). (From DeFeudis and Paolino, 1972)

*Cerebral Glycoproteins*

Considering the studies of Bogoch (1968, 1972, 1973), it seems possible that some of the radioactivity incorporated into the brain after subcutaneous injections of labeled monosaccharides (e.g., glucose; see above) could have been related to the synthesis of brain glycoproteins. Glycoproteins are present in high concentration in cerebral nerve ending membranes (Quamina and Bogoch, 1968; Gombos et al., 1972), and their concentrations change with increasing developmental complexity of the nervous system (Bogoch, 1968), and during regression of higher brain functions (Bogoch, 1962; Bogoch and Belval, 1966). Some synaptic receptors and neurotransmitter enzymes might be glycoproteins (see DeFeudis and DeFeudis, 1977). Certain cerebral glycoprotein fractions were increased by operant conditioning (training) of pigeons (Bogoch, 1965, 1968, 1973). After intravenous injections of $1^{14}C$-D-glucose into pigeons, only a slight incorporation of radioactivity occurred into cerebral glycoproteins at rest, but significantly more occurred after 10 minutes, or longer, of training. Other results have indicated that the absolute amount of "glycoprotein 11A" in the brain was correlated with the "amount learned" by pigeons (see Bogoch, 1968, 1972, 1973). Hence, these substances could be involved in the "coding" of experiential information.

In general, the studies of Bogoch (1968, 1972) have revealed that the cerebral concentration of protein-bound hexose increases with increased complexity of cerebral anatomical structure and with increased behavioral complexity or activity of the animal. These studies were corroborated to some extent by those of Ramirez et al. (1974), who showed that incorporation of the glycoprotein precursor, glucosamine, into rat brain regions was affected by environmental complexity. After 30 days of differential housing, intraventricularly-administered $^3H$-glucosamine was incorporated to a greater extent into subcellular fractions of both the cerebral cortices and hippocampi of "enriched" rats at one hour after injection. At three hours after injection, more label was found in subcellular fractions of the "deprived" rats, except for hippocampal plasma membrane which was still more heavily labeled in "enriched" rats. These results are in accord with other findings which have indicated that the hippocampus plays an important role in information processing.

## Putative Neurotransmitter Systems

### *Introduction*

Centrally-active amino acids and acetylcholine (ACh) are formed from glucose in the brain. Upon injection of labeled precursors of acetyl-CoA into mammals, labeled ACh appears rapidly in the brain (Cheney et al., 1969; Tucek and Cheng, 1970). Carbon atoms of systemically-injected U-$^{14}$C-D-glucose are rapidly incorporated into active amino acids of the brain (Vrba, 1962), and these amino acids can be both synthesized in and released from various cerebral structures (DeFeudis et al., 1969, 1970). Since ACh and active amino acids are formed from glucose, approaches using labeled glucose injections with subsequent analysis of newly-formed transmitter suspects will yield results which may link cerebral neurotransmitter mechanisms and energy metabolism to environmentally-induced behavioral modification.

### *Central Cholinergic Systems*

Both total AChE and ChE (pseudo-ChE) activities of the cerebral cortices (especially the occipital cortices) of rats were increased by "enriched" environments. However, expression of these results on a unit weight basis revealed that AChE was *decreased* by environmental enrichment, whereas ChE was increased. Thus, the increased cortical growth caused by enriched experience could have occurred mainly in elements which are low in AChE activity (e.g., glia, non cholinergic neurones). The occipital cortices of $A_1$ and C57 strains of mice and of gerbils also contained less AChE per unit weight when these animals were reared under "enriched" conditions (Rosenzweig and Bennett, 1969). On the other hand, Goldberg and Welch (1972) found that neither choline acetylase (ChAc) nor ChE activity was altered in the brains of socially stimulated mice that had been subjected to isolation-rearing; these results agreed with those of Consolo and Valzelli (1970), who used the same assay method for ChAc. However, these results are not strictly comparable since Goldberg and Welch (1972) used mice that had been individually caged, but intermittently grouped, during the study.

Using differentially-housed CF-1 mice, Essman (1971a) showed that both AChE and ChE activities were lower in the brains of "aggregated" mice, and that environmental stimulation of "isolated" mice significantly elevated the activities of both enzymes in somatosensory cortex. Perhaps Essman's most interesting finding was that the *direction* of the effect of environmental stimulation was contingent upon the endogenous condition caused by differential housing. In somatosensory cortex, decreases in cholinesterases were shown in "aggregated" mice given environmental stimulation, whereas increases were shown in "isolated" mice subjected to the same stimulation. In these studies, 21-23-day-old mice were housed for 28 days, either singly or in groups of five (see also Essman, 1972, and Proteins Section). The difference between these results and those of the Berkeley group (Bennett and Rosenzweig, 1971) is probably related mainly to the different housing conditions employed.

*γ-Aminobutyric Acid (GABA) and Other Amino Acids*

Increases in GABA and glutamate in thalamus-hypothalamus, and decreases in caudate nucleus were produced by keeping mongrel puppies under "partial isolation" for a week (Agrawal et al., 1967; see also Himwich, 1969). Less GABA was bound to cerebral subcellular fractions, containing nerve endings, when C57 male mice were kept "isolated" than when they were kept "aggregated" for 35-37 days, beginning on day 21 (DeFeudis, 1972d). This finding agreed with the suggestion of Bennett et al. (1964) that cerebral changes caused by experience could occur mainly in non-cholinergic elements and glial cells; that is, in elements which were expected to be rich in GABA receptors and uptake sites. The "binding" of GABA appeared to provide a measure of cerebral nerve endings which could be affected by chronic environmental changes (see also DeFeudis et al., 1976a). Thus, animals exposed to isolation, by having a lower capacity to bind GABA (an inhibitory substance), might have less inhibitory nerve endings. This might be related to the behavioral changes (e.g., increased arousability and aggressiveness) caused by isolation. These results are in accord with those of Lal et al. (1972), who showed that the brains of male, Swiss albino mice, "isolated" for 6 or 20 weeks, contained less GABA than those kept in groups of ten. Earley and Leonard (1977) also found lower levels of GABA in "isolated" mice, as compared to grouped mice. The tendency of mice to exhibit aggressive responses appeared to be inversely related to GABA contents in certain brain regions.

*Norepinephrine (NE), Dopamine (DA) and Serotonin (5-HT)*

*Introduction.* Catecholamines and 5-HT appear to be involved in human affective states, and in the regulation of aggressiveness, stereotypies, sleep states, and learning ability. Mouse strains possessing lower levels of cerebral catecholamines and 5-HT generally appear to be more aggressive and to exhibit less stereotyped behavior than strains having higher levels of these substances (Maas, 1962; Bourgault et al., 1963). Agressiveness might be elicited by an increase of cerebral NE, whereas stereotypy might be caused by increased cerebral DA (Randrup and Munkvad, 1968, 1970). Aggressive behaviors in rats and cats might also be associated with increases in cerebral NE (Hernandez-Peon et al., 1963; Reis and Fuxe, 1969; Fog et al., 1970).

*Cerebral concentrations of NE, DA, and 5-HT.* Welch (1965) showed that male DUB/ICR mice, living as a group of 20 for a week, had lower cerebral NE concentrations than mice which had been individually housed for the same period of time. However, Welch and Welch (1968a, 1968c) found that cerebral NE content was higher in "grouped" than in "isolated" male mice, and Valzelli (1967a) found that cerebral catecholamine concentrations of male mice were not significantly modified by isolation. In other studies, Essman (1971b) showed that cerebral NE concentration did not differ in CF-1 mice that had been kept "isolated or "grouped" for 28 days, commencing at 21 days of age. However, when differential housing had been carried out for 34 days, "isolated" mice had lower cerebral NE than "grouped" mice (Essman and Smith, 1967). "Isolated" rats (weaned at 18 days of age) possessed higher levels of cerebral NE than group "enriched" rats (Geller et al., 1965), an increase which appeared to be localized in the caudate nucleus (Geller and Yuwiler, 1968). As all possibilities are covered by these results (increase, decrease, no effect), it is not yet possible to state the relationship between differential housing and cerebral NE content.

Garattini et al. (1969) found no change in the 5-HT contents of either whole brain or discrete cerebral structures in differentially housed mice (see also Valzelli, 1967a), whereas Welch and Welch (1968a, 1968c) found higher 5-HT contents in the telencephalon and diencephalon–mesencephalon of "grouped" than of "isolated" mice. Isolation of mice caused a decrease in cerebral 5-hydroxyindoleacetic acid (5-HIAA) content within one day, and this change persisted for at least a month (Giacalone et al., 1968). In albino CF-1 mice, "isolated" for 34 days (commencing at 21 days of age), Essman and Smith (1967) could find no change in either whole brain 5-HT or 5-HIAA, but subsequent results (Essman, 1968, 1969) indicated that isolation of this same mouse strain for 24 days produced an increase in whole brain and regional 5-HT and 5-HIAA. From these studies, no firm conclusions can be drawn with respect to environmental effects on 5-HT content. The discrepancies may be mainly due to variations in experimental design (e.g., duration of differential housing, differences in the time of onset of isolation, and strain differences). It must also be remembered that concentration levels reflect a dynamic homeostasis between production and breakdown, so that turnover rates may be modified without being detectable by measuring concentration.

*Cerebral turnover of NE, DA and 5-HT.* "Isolated" male mice had lower rates of synthesis, release and turnover of cerebral catecholamines than did "grouped" mice, as indicated by a greater decrease of catecholamines in "grouped" mice after administration of the tyrosine hydroxylase inhibitor, dl-α-methyltyrosine (α-MT) (Welch, 1967; Welch and Welch, 1968c). Impairment of catecholamine synthesis by a α-MT markedly decreased agressiveness in "isolated" mice (see Welch and Welch, 1969a). Accumulation of cerebral 5-HT and catecholamines was usually slower following administration of monoamine oxidase inhibitors (MAOI) to "isolated" mice (Welch and Welch, 1966, 1971; Garattini et al., 1967; Giacalone et al., 1968; Valzelli and Garattini, 1968). These differences in the rates of cerebral NE and DA turnover appeared to be graded with respect to level of environmental stimulation (group size), and graded differences in the accumulation of these brain amines were shown within a few hours after injection of the quick-acting MAOI, pargyline (see Welch and Welch, 1971). Valzelli (1971) showed a minor reduction of NE turnover and an increase in DA turnover in the brains of "isolated" mice. (Table 5). Again, these disparities perhaps reflect mainly technical differences, as well as strain and age differences. Cerebral NE and DA contents were decreased by para-chlorophenylalanine (p-CPA, an inhibitor of 5-HT synthesis) to about the same small degree as 5-HT; these changes were more pronounced in "grouped" than in "isolated" mice (Welch and Welch, 1968a).

Brain 5-HT was increased more in "grouped" than in "isolated" mice after administration of the MAOIs, pargyline (Welch and Welch, 1971) and tranylcypromine (Valzelli, 1967a; Garattini et al., 1967). "Isolated" mice showed a decreased 5-HT turnover rate in brain and a concomitant increased 5-HT turnover time when compared to "grouped" (10 per cage) mice (Giacalone et al., 1968). These differences were also evident in brain regions, being greatest in the corpora quadrigemina, in which the rate was only two-thirds as fast in "isolated" as in "grouped" mice. Another report revealed that brain 5-HT was elevated, and that brain 5-HT turnover was increased in "isolated" CF-1 and Swiss-Webster mice, as compared to their "grouped" (5 per cage) counterparts (Essman and Smith, 1967; Essman, 1968). These results differed from those reported by Welch's and Garattini's groups, perhaps due to the earlier age (21 days) of initiation of differential housing, or due to the taking of brain samples immediately after mice had been tested for behavioral activity in a novel situation. Therefore, the basal rate of metabolism was not determined in the latter studies.

In general, the results of Welch's and Garattini's groups are in agreement. 5-HT turnover occurs more rapidly in "grouped" mice, and the rate of synthesis of 5-HT increases following tranylcypromine administration. The decrease in 5-HT turnover represents the most typical biochemical change in monoamine systems which occurs in "isolated," aggressive mice (see Valzelli 1973a; Table 5). However, it should be noted that naturally high aggressive rats exhibited faster 5-HT turnover than low aggressive rats in two brain regions (Daruna and Kent. 1976).

*Other factors affecting cerebral levels, metabolism, and turnover of NE, DA and 5-HT.* Many factors affect the metabolism of these possible meurotransmitters. Although these make it more difficult to control experiments designed to quantitate environmentally induced changes in monoamine systems, they have been useful for sensitizing the methods used. A good example is provided by the results obtained using drugs that block the rate-limiting enzymes of catecholamine and 5-HT synthesis (α-MT and pCPA). These agents, when administered to mice, enhanced the effects of differential housing on the cerebral metabolism of these proposed transmitters. In some cases in which equivocal results were obtained in the absence of drugs (e.g., for cerebral contents of NE, DA, and 5-HT), the use of the drugs, restraint stress, or fighting activity accentuated the differences caused by differential housing.

Along this line, Welch and Welch (1969b), using DUB/ICR male mice that had become aggressive after individual housing, showed that daily fighting (for 5-10 minutes) for either 10 or 14 days caused a significant increase in the whole brain DA levels; cerebral NE was elevated to the same extent, but only after 14 days; no change occurred in cerebral 5-HT (Fig. 2). This finding agreed with others which indicated that cerebral NE content was elevated in rats subjected to electro-convulsive shock twice a day for seven days (Kety et al., 1967), or to three days of repetitive stress (Thierry et al., 1968). Such sustained effects can differ (*in direction*) from acute effects; for example, the acute effect of intense fighting is a lowering of brain NE and DA, at least in the brainstem (Welch and Welch, 1968d). These sustained effects on catecholamines perhaps reflect adaptive mechanisms. Welch and Welch (1969b)

Table 5. Cerebral Serotonin (5-HT), Norepinephrine (NE) and Dopamine (DA) Turnovers in Normal and "Isolated" Mice

| | 5-HT turnover | | NE turnover | | DA turnover | |
|---|---|---|---|---|---|---|
| Type of mice | Rate (μg/g/hr) | Time (hr) | Rate (μg/g/hr) | Time (hr) | Rate (μg/g/hr) | Time (hr) |
| Normal | 0.46 | 1.10' | 0.05 | 8.32' | 0.08 | 6.10' |
| "Isolated-aggressive" | 0.26[a] | 2.20' | 0.03[a] | 10.15' | 0.13[a] | 3.54' |

Means of 8 determinations; [a], indicates $p < 0.01$ with respect to normal animals.
(From Valzelli, 1971).

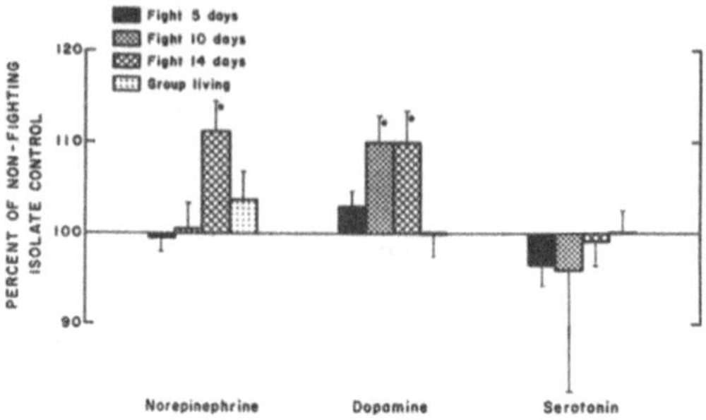


Fig. 2. Effect upon whole brain catecholamines of permitting individually housed male mice to fight for 5-10 min. daily as compared with undisturbed group-housed and individually housed mice. Each mean and SEM is presented as a percentage of the non-fighting isolate control mean. Black dots indicate individual means that differ significantly from controls (at least $p < 0.05$). (From Welch and Welch, 1969b).

found that the decrease in cerebral catecholamines caused by fighting occurred first in brain stem, the marked decreases in pons-medulla being of interest since cell bodies containing these amines are localized in this area (Hillarp et al., 1966). Dopamine was reduced to about 59% of its control value in the brain stems of mice that fought in groups of 4 for 150 minutes, whereas no change occurred in the telencephalon (Welch and Welch, 1971). With "isolated" mice that were exposed to fighting activity by suspending them in a cage over 4-6 other "isolated" mice, it was shown that merely witnessing fighting for 75 minutes could markedly reduce the NE content of the pons-medulla (Table 6; Welch and Welch, 1968b). Hence, a purely psychological stimulus appeared to enhance cerebral NE release (c.f., studies on environmental enrichment by Ferchmin et al., 1975).

More recently, Hendley et al. (1973) showed that the $K_m$ for NE uptake into cerebral cortical homogenates and the $K_i$ for the d-amphetamine-induced inhibition of this uptake were increased in "grouped" mice which

Table 6. Reduction of Brain Norepinephrine in Male Mice That Had Witnessed Fighting Among Other Male Mice Made Aggressive by Long-Term Isolation*

| Brain region | Norepinephrine (ng/g) Undisturbed controls | Strange cage | Witness fighting |
|---|---|---|---|
| Telencephalon | 568 ± 17.4 | 557 ± 11.4 | 531 ± 17.0 |
| Diencephalon and mesencephalon | 524 ± 25.2 | 583 ± 20.5 | 522 ± 26.6 |
| Pons and medulla oblongata | 472 ± 24.9 | 421 ± 22.4 | 319 ± 23.4+ |

* Values are expressed as mean ± S.E.M.; n=20 in all cases.
\+ Differs $p < 0.001$ from controls, $p < 0.002$ from strange cage (Student's t-test; two-tailed).
Mice were placed in a feed basket above an empty strange cage or above a strange cage that contained fighting mice for 75 minutes. All brain parts were analyzed separately for individual animals.
(From Welch and Welch, 1968b)

had been living under chronic attack from aggressive cagemates. $K_m$ and $V_{max}$ for NE uptake were also increased 18-20 hours after the last of a series of short, intense, daily fights between male mice that had been made aggressive by prior long-term individual housing (see also Welch et al., 1974b). These results revealed that the cerebral NE uptake mechanism might be adaptable and plastic. The elevated $V_{max}$ caused by fighting indicated that an increase had occurred in the number of NE uptake sites on neuronal membranes, and the elevation in $K_m$ could indicate a decreased affinity of NE for its uptake sites.

There is little doubt that circadian, ultradian and seasonal rhythms affect cerebral neurotransmitter metabolism. A phase relationship between NE and 5-HT contents of rat brain has been noted (Friedman and Walker, 1968). 5-HT peaked during the light phase and was associated with periods of inactivity and sleep, whereas NE peaked during the dark phase and was associated with increased motor activity. Both circadian and seasonal rhythms have been observed for cerebral 5-HT and 5-HIAA in mice by Garattini and Valzelli (1965). Recent findings which revealed that the toxicity (Walker, 1972), and the effects of amphetamine on locomotor activity of mice (del Pozo et al., 1978), are influenced by time of day as well as by group size, might be related to fluctuations in endogenous cerebral catecholamine content. Extent of handling, age at sacrifice, age of weaning, and sex and strain of animals can also alter cerebral biogenic amines and other substances.

## EFFECTS OF ENVIRONMENT ON THE CENTRAL ACTIONS OF PSYCHO-ACTIVE AGENTS

### *Amphetamines*

The dextro isomer of amphetamine (d-amphetamine) has perhaps been the most widely employed psycho-active agent in studies on differentially housed animals. It seems necessary to emphasize that *two different types* of environmental studies have been performed with this agent (see below).

### *"Short-Term" Environmental Effects*

"Acutely-grouped" mice are more sensitive to injected sympathomimetic amines (e.g., d-amphetamine) than "acutely isolated" mice (Gunn and Gurd, 1940; Chance, 1946). This is a *short-term* effect, since in such studies mice were housed in colonies prior to experiments, treated with the drug, and then differentially housed only during the period following treatment. Short periods of aggregation potentiate both the behavioral excitation and the lethality of injected d-amphetamine in mice (e.g., Chance, 1946; Weaver and Kerley, 1962; Moore et al., 1965). The increased toxicity of d-amphetamine caused by aggregation of mice is related to its production of hyperactivity, hyperthermia, hypoglycemia and/or cardiovascular damage, as well as to CNS damage. The most important action is perhaps the hyperthermia-induced CNS damage (e.g., Clark et al., 1967; Welch and Welch, 1973). The hypoglycemia is caused perhaps by the marked increase in glucose utilization which is promoted by the drug-induced hyperactivity (Bewsher et al., 1966). The important point which emerged from these studies is that even "short-term" modification of social environment can cause pronounced effects on an animal's responsiveness to CNS stimulants.

### *Effects of "Prolonged" Exposure to Differential Housing*

The writer will emphasize studies which have been performed on animals subjected to "prolonged" periods (several weeks) of differential housing, since such "long-term" changes might be more significant in explaining the environmental bases of certain human problems such as emotional changes and drug addictions.

In this type of study, d-amphetamine is more toxic to "isolated mice (e.g., Consolo et al., 1965a; Welch, 1965; Welch and Welch, 1966, 1971). Welch and Welch (1971) used DUB/ICR mice which had been differentially housed for five weeks, injected with 45 mg/kg d-amphetamine (a lethal dose in this strain), and then placed into individual cages. The mean dying time of "grouped" mice was nearly twice that of "isolated mice, and "isolated" mice reacted more uniformly to the drug than did "grouped" mice. The greater variability shown among the "grouped" mice was explained by previous findings which revealed that the subordinate mice of a group hierarchy were very insensitive to this drug, and that dominant (more aggressive) mice reacted similarly to "isolated" mice (see Welch and Welch, 1966). Another interesting point brought out by Welch and Welch (1966) was that regardless of their prior housing conditions, mice that were grouped subsequent to d-amphetamine injection died more quickly than those placed in isolation after drug treatment (Table 7; see Gunn and Gurd, 1940; Chance, 1947). This "short-term" reversal of the effects of prior housing conditions on the action of d-amphetamine has been demonstrated on a metabolic level by monitoring the effects of d-amphetamine on the entry of glucose or mannose carbon atoms into the brains of mice (DeFeudis et al., 1972; DeFeudis, 1973). Welch and Welch (1966) showed further that the effect of social

Table 7. Effects of Pre-Drug and Post-Drug Housing Conditions upon d-Amphetamine Toxicity

| Dose of d-Amphetamine Sulfate (mg/kg) | Time to Die (min.) Pre-isolated Post-isolated | Pre-isolated Post-grouped | Pre-grouped Post-isolated | Pre-grouped Post-grouped | p |
|---|---|---|---|---|---|
| 25 | 73.0 ± 6.4+ | 10.4 ± 4.9 | 84.5 ± 5.9 | 62.2 ± 5.8 | |
| 35 | 69.1 ± 10.1 | 21.2 ± 6.1 | 106.0 ± 17.2 | 61.7 ± 7.5 | |
| 45 | 36.9 ± 2.1 | | 72.6 ± 17.7** | 50.9 ± 15.7 | |
| 75 | 34.3 ± 1.8 | 12.0 ± 3.8 | 76.0 ± 20.8 | 65.0 ± 27.6 | |
| 90 | 26.2 ± 1.8 | 10.5 ± 2.5 | 60.2 ± 29.5 | 22.5 ± 8.3 | |
| 130 | 8.1 ± 0.8 | | 77.7 ± 41.3 | | |
| 140 | 8.6 ± 1.6 | | 7.8 ± 2.1 | | |
| Mean++ | 36.6 | | 64.3 | | .01 |
| | | 13.6 | | 52.9* | .00 |
| | | | 69.3 | 52.5 | .05 |
| | 50.7 | 13.6 | | | .00 |

* analysis of variance, method of adjusted means for nonhomogeneous variance
\+ standard error of the mean.
** Plus one survivor.
++ Means computed for each comparison are based upon relevant common dosages only.
Male DUB/ICR mice were kept in isolation or in groups of 5 for five weeks in opaque plastic cages with superabundant food and water, after which they were given a lethal dose of amphetamine sulfate (i.p. in 0.1 ml of 0.9% saline) and isolated or grouped to die. Each mean dying time represents 10 animals, except for the pre-isolated post-grouped animals, for which each mean represents 20 animals (From Welch and Welch, 1966).

stimulation upon d-amphetamine toxicity was inversely related to the number of mice living together in a "graded" series which included "isolated" animals and groups of 2, 5 and 10. Both spontaneous aggressiveness within groups and aggressiveness toward strange mice were decreased by increasing the group size of prior housing, effects which paralleled the reduction in CNS excitability, as indicated by d-amphetamine. Lal et al. (1972), using randomly bred, Swiss albino mice, housed either individually or 10 per cage, showed that the stimulatory effects of amphetamine on both exploratory activity and spontaneous locomotion were enhanced by prior individual housing of the mice.

Welch and Welch (1973) have defined long-term group living of mice as a "chronic stress" which increases their tolerance to d-amphetamine. Another stress, "sleep deprivation," also increased tolerance to amphetamine (Ferguson and Dement, 1969). Hence, the decreased sensitivity of mice to d-amphetamine, produced by group living, appears to be paralleled by a general reduction of CNS responsivity to stimuli.

Using weanling, male, Swiss albino mice, housed in groups of either 5 or 10, it has been shown that mice which frequently attacked other mice but were not themselves attacked (i.e., "aggressors") were more alert and generally more responsive to environmental stimuli than mice which were frequently receivers of attack but never observed to attack another mouse (i.e., "receivers"; Welch, 1967; Welch and Welch. 1973). Aggressors were also much more sensitive than receivers to the toxic effects of injected d-amphetamine. Among mice that lived in groups, the period of survival after injection of d-amphetamine (110 mg/kg, i.p.) was nearly an hour less for aggressors than for receivers, and after administration of pargyline, more NE and DA accumulated in the brains of receivers than in those of aggressors (Welch and Welch, 1973). Hence, tolerance to d-amphetamine increased as the number of mice living together was increased, and as the frequency with which individual mice were involved in fighting was decreased (Welch and Welch, 1973). This finding is supported to some extent by the findings that prior electric shock reduced mortality from acute electric shock and protected against brain NE depletion (Zigmond and Harvey, 1970).

These results are also supported by recent findings which showed that d-amphetamine (50 mg/kg, i.p.) was removed from the brain more rapidly in "grouped" than in "isolated" male C57 Black mice, suggesting less uptake by the brains of grouped mice. Also, less $^3$H of injected $^3$H-d-amphetamine was bound to a region of linear sucrose gradients containing synaptic membranes and vesicles prepared from the brains of grouped mice (DeFeudis and Marks, 1973a; see Amphetamines Section). Hence, the greater toxicity of d-amphetamine in "isolated" mice appears to be correlated with its greater retention by the brain. d-Amphetamine and methamphetamine (which is about twice as potent on a unit weight basis as d-amphetamine) have also been shown to enhance the effects of "enriched" environmental conditions on brain weight and cholinesterases in rats (Rosenzweig and Bennett, 1972; Bennett et al., 1973). Therefore, these sympathomimetic amines might produce an increase in arousal level which mimics, at the level of central synapses, the effects of environmental "enrichment", or "grouping", of animals.

In general, the concepts of "arousability" and "arousal" (see Walsh and Cummins, 1975) are most useful for explaining the different actions of amphetamines. "Short-term" isolates, which have low arousal (due to little stimulation) and normal arousability (since there is a lag period before homeostasis is effective), will be less affected by amphetamines. On the other hand, "long-term" isolates, which have a high level of arousability, will be more affected by amphetamines.

The relevancy and possible application of these studies to human situations is obvious, since environment, and the degree of enrichment derived from it, could be major factors involved in drug addiction.

Sedative Hypnotics and Ethanol

Stimulus deprivation of mice appears to reduce their sensitivity to CNS depressants, while increasing their sensitivity to CNS stimulants (Lal et al., 1972). Using sedative-hypnotics (hexobarbital, pentobarbital, chloral hydrate and barbital), the duration of drug-induced narcosis was shown to be markedly reduced by exposure of Swiss albino mice to individual housing (Baumel et al., 1970; Lal et al., 1972). The magnitude of this decrease in narcosis duration was greatest for hexobarbital. This change in sensitivity to sedative hypnotics did not appear to be related to the development of aggressiveness (Baumel et al., 1970). Both male and female mice reared in isolation showed a more rapid decline of whole body concentration of hexobarbital, which indicated enchanced hepatic metabolism of this drug (Baumel et al., 1970), and both recovered from narcosis at hexobarbital concentrations which were markedly higher than those of "grouped" mice (Lal et al., 1972). As control for this hepatic effect, the duration of action of barbital (an agent which is not a substrate for hepatic drug-metabolizing enzymes; Dorfman and Goldbaum, 1947) was shown to be shorter in "isolated" mice (Baumel et al., 1969). These findings support the notion that the CNS of "isolated" mice is less sensitive to depressant drugs, perhaps due to their higher level of arousability, and should be contrasted with those of Welch and Welch (1966, 1973), which indicated that "grouped" mice displayed increased tolerance to the excitant drug, d-amphetamine, and with those of Yanai and Ginsburg (1976), which revealed that individually housed mice were more sensitive to the effects of chronic ethanol treatment.

Major and Minor Tranquilizers

Major tranquilizers (reserpine, chlorpromazine [CPZ]) exerted greater effects upon "grouped" than "isolated" animals (e.g., Consolo et al., 1965b). The effect of CPZ - reducing a previously learned avoidance response - was less pronounced in "isolated mice (Lal et al., 1972), thus supporting the idea of reduced sensitivity of "isolated" mice to CNS depressants. Regarding "minor" tranquilizers ("anti-anxiety" agents), both aggressive behavior and impairment of exploratory activity have been shown to be particularly sensitive to the action of benzodiazepines (Valzelli et al., 1967, 1971; Valzelli and Bernasconi, 1971; Valzelli, 1973a, 1973b, 1973c). Also, medazepam constantly improved learning performance in "isolated" mice, while reversing the alteration in cerebral 5-HT turnover in these animals (see Valzelli, 1973a).

Morphine

Few studies have been performed on the possible environmental effects on morphine action. Regarding "short-term" changes in environment, Spoerlein (1968) reported that the $LD_{50}$ of morphine differed between "isolated" and "aggregated" mice, but this was not confirmed (Vedernikov, 1970), and Davis and Brister (1971) showed that the lethality of injected morphine and other narcotic analgesics was greater in "aggregated" mice. "Long-term" isolation increased the sensitivity of hooded rats to morphine-induced analgesia (Katz and Steinberg, 1972), but this effect was not observed in another rat strain (Adler et al., 1975b). However, Adler et al. (1975a) did show that differential housing could alter the abstinence syndrome in morphine-dependent rats that had been previously subjected to long-term isolation. More recently, male, Swiss albino mice, weaned at 21-22 days of age and subjected to differential housing for 10-13 weeks were used, and it was shown that "isolated" mice displayed less sensitivity to pain (hot-plate test) than did their "aggregated" counterparts after injection of morphine (DeFeudis et al., 1976b, 1978). Thus, animals subjected to isolation-rearing appear to be more susceptible to the effect of acutely administered morphine (see also Kostowski et al., 1977). Changes in the cerebral concentrations or metabolism of endogenous morphine-like substances (enkephalins; see Hughes, 1975; Terenius and Wahlstrom, 1975) could be involved in mediating such changes.

Comment

The above-mentioned studies have revealed that *environment alters the actions of psycho-active agents*. This effect of environment involves whole body metabolism rather than exclusively that of the brain.

## EFFECTS OF ENVIRONMENT ON THE METABOLIC DISPOSITIONS AND METABOLIC ACTIONS OF PSYCHO-ACTIVE AGENTS

### Introduction

This new area of research could be important in solving human problems related to changes in CNS excitability and behavior that are caused by changes in social environment.

### Changes in Metabolic Dispositions of Drugs

#### *Barbiturates*

The increased CNS excitability of "isolated" mice appears to be mediated, in part, by a change in metabolism which increases the rate of disappearance of hexobarbital from the body (see page 90; Sedative Hypnotics and Ethanol). Since individual housing may activate the pituitary-gonadal axis (Brain and Nowell, 1969, the enhanced hepatic metabolism shown in these studies may be due to an initial hyperactivity of the posterior hypothalamus, an area which is believed to regulate heoatic drug-metabolizing enzymes (Nair et al., 1969).

#### *d-Amphetamine*

Thirty minutes after intraperitoneal injection of $^3$H-d-amphetamine to differentially housed male C57 Black mice, more radioactivity was present in both the brains and sera of "isolated" mice (DeFeudis and Marks, 1973b). The time-course of entry of $^3$H (of injected $^3$H-d-amphetamine) into the brain revealed that during the declining phase of cerebral radioactivity (approximately 15 - 120 minutes post-injection) the brains of "isolated" mice contained more of the injected drug (+ metabolites) than those of their "grouped" littermates (DeFeudis and Marks, 1973a). Hence, it appeared that this stimulant drug was metabolized and excreted less rapidly in "isolated" mice, a finding which might be related to the greater toxicity of d-amphetamine in "isolated" mice. The different rate of decline of $^3$H between the brains of "isolated" and "grouped" mice was likely due to metabolites of the injected drug. These latter studies lead one to suspect that the potential for drug addiction may be related to the state of arousal of the animal (or level of excitability of the brain) at the time during which the drug is taken. An organism's state of arousal should be expected to influence its capacities for retention, metabolism and excretion of potentially addictive drugs (see below).

### Effects of Drugs on Energy Metabolism

Results presented above (Carbohydrate Metabolism section) indicated that carbon atoms of subcutaneously injected U-$^{14}$C-glucose, or other substrates of energy metabolism, were incorporated to a greater extent into the brains of "grouped," than of "isolated" male mice. Since this difference was monitored in the brain, it seemed likely that environmentally induced differences in the actions and metabolism of psycho-active drugs would also be reflected in carbon atom incorporation into the brain.

In line with this rationale, pretreatment of mice with LiCl (200 or 400 mg/kg, i.p., 3 hours prior to subcutaneous administration of 1-$^{14}$C-pyruvate) increased the incorporation of pyruvate carbon atoms into the brain and liver by an action which was selective to "isolated" mice "DeFeudis and Paolino, 1972; Table 4). This effect could have been mediated by an increased retention of $Li^+$ by subcellular structures of the brains of "isolated" mice (DeFeudis, 1972f; see below). Other studies revealed that d-amphetamine, injected 30 minutes before "tracer" U-$^{14}$C-D-glucose, selectively reduced carbon atom incorporation into the brains of "isolated" mice (DeFeudis et al., 1972; DeFeudis, 1973). Thus, both $Li^+$ and d-amphetamine acted selectively on the metabolism of the "isolated" mice, and glucose carbon atom incorporation into the brain was changed in opposite directions by these two agents, which exert opposite actions on CNS excitability.

In further experiments, male, weanling, Swiss-Webster mice, kept singly or in groups of 25 for 3.5 weeks, were treated with either d-amphetamine-$SO_4$ (10 mg/kg) or the vehicle (0.154 M NaCl) intraperitoneally, one hour before sacrifice, and with a subcutaneous "tracer" dose of 1-$^{14}$C-D-mannose, 30 minutes before sacrifice (DeFeudis, 1973). As shown with the C57 Black strain, carbon atom incorporation occurred to a greater extent into the brains of "grouped" mice. d-Amphetamine decreased the entry of mannose carbon atoms into the brains of "isolated" mice, and appeared to increase this slightly in "aggregrated" mice. The major difference in the action of d-amphetamine on cerebral incorporation of radioactivity was found between "isolated" and "grouped" mice housed under conditions similar to their prior housing conditions during experiments. In these cases, injected d-amphetamine markedly decreased radioactivity of the sera of "isolated" mice, which were grouped for 1.5 hours during experiments, whereas it exerted no such effect on mice that were kept "isolated" during experiments. In "grouped" mice, d-amphetamine decreased serum radioactivity only when animals were kept in groups of 10 during experiments. Brain:serum ratios of $^{14}$C revealed that the difference between carbohydrate metabolism of "isolated" and "grouped" mice was made more significant by injecting d-amphetamine. Although many factors are involved in explaining these findings (e.g., changes in hormone levels, motor activity, body temperature, cellular permeability, circulation), it seems evident that a metabolic basis exists for environmentally induced changes in the action of d-amphetamine.

## EFFECTS OF ENVIRONMENT ON THE CEREBRAL SUBCELLULAR LOCALIZATIONS OF PSYCHO-ACTIVE AGENTS

### d-Amphetamine

Studies on the cerebral subcellular distribution of $^3$H-d-amphetamine revealed both *in vivo* (after injection of drug) and *in vitro* (drug added to homogenizing fluid) that extensive "binding" (or retention) of this drug (+ metabolities, *in vivo*) occurred to brain particles. This "binding" was more pronounced in a fraction which was enriched in nerve endings than in a fraction which was enriched in nuclei and cellular debris, and it occurred to a greater extent in the brains of "isolated" mice when the drug was administered *in vivo*. Further analysis revealed that the particles involved in the "binding" may have been fragments of nerve ending membranes and/or synaptic vesicles (DeFeudis and Marks, 1973a; Fig. 3).

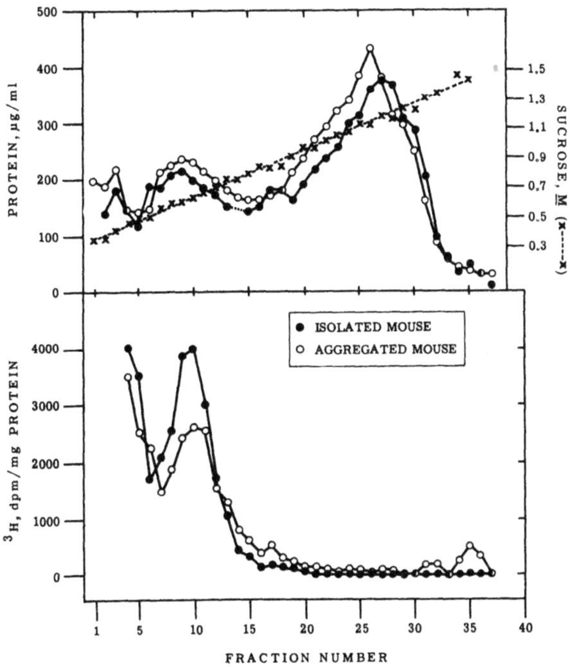


Fig. 3. Subcellular profiles of the $^{3}H$ and protein contents of synaptosome-enriched "$P_2$" fractions prepared from the brains of "isolated" and "aggregated" C-57 Black mice. Both mice were housed differentially for 5 weeks. About 17 min. after injection of $^{3}H$-d-amphetamine-$SO_4$ (50 mg/kg; 0.1 µCi/g), the mice were decapitated and their brains were homogenized in 10 vols 0.32 M sucrose solution. Synaptosomal fractions were re-suspended in 3.5 ml 0.32 M sucrose solution, and 2.0 ml of each sample were layered onto linear gradients (0.4-1.8 M sucrose) and centrifuged at 137,000 × g, 1.5 hr. Small fractions of the gradients were collected using a gradient fractionator and analyzed for their contents of $^{3}H$ and protein. Upper graphs indicate the subcellular distributions of total protein and the linearity of the gradients; Lower graphs indicate the difference in the incorporation (or "binding") of $^{3}H$ which existed between the brains of the "isolated" and the "aggregated" mouse, as well as the localization of $^{3}H$ to regions of the gradients which contained 0.5-0.7 M sucrose. (From DeFeudis and Marks, 1973a).

## Lithium

With both low (4 mM) and high (40 mM) concentrations of LiCl in homogenizing fluids, a significantly greater percentage of the total $Li^+$ was bound to subcellular fractions prepared from the brains of "isolated" mice, as compared with those of their "grouped" littermates (DeFeudis, 1972f; Table 8). Hence, by merely adding $Li^+$ to homogenizing fluids, changes in cerebral subcellular morphology caused by differential housing were demonstrated.

Table 8. Retention of $Li^+$ by Particulate Fractions of the Brains of Differentially Housed Mice

| Salts in homogenizing fluid | $P_1$ fraction | | $P_2$ fraction | |
|---|---|---|---|---|
| | μEq $Li^+$/g protein | % $Li^+$ | μEq $Li^+$/g protein | % $Li^+$ |
| "Isolated" mice | | | | |
| 4mM NaCl + 4mM LiCl | 78.4 ± 1.1 | 15.2 ± 1.8* | 57.3 ± 3.1 | 15.7 ± 0.7+ |
| 4mM NaCl + 40mM LiCl | 591.0 ± 4.9** | 19.9 ± 0.4* | 745.0 ± 39.2* | 5.4 ± 0.4+ |
| "Aggregated" mice | | | | |
| 4mM NaCl + 4mM LiCl | 75.5 ± 6.9 | 11.9 ± 1.0 | 57.8 ± 1.0 | 14.3 ± 0.6 |
| 40mM NaCl + 40mM LiCl | 468.0 ± 20.0 | 17.4 ± 0.8 | 590.7 ± 42.1 | 3.7 ± 0.1 |

Weanling, male C57 Black mice were housed either singly, or in groups of 25, for 35-37 days; all animals were food-deprived for 19-21 hours prior to their sacrifice.

Means ± standard errors; 5 mice in all cases. *, +, and ** indicate, respectively, $P < 0.05$, $p < 0.01$ and $p < 0.001$ for comparisons between these values and the corresponding values for "aggregated" mice.

$P_1$ = pellet fraction from 1000 x g, 10 min, 0°C centrifugation. $P_2$ = pellet from 17,500 x g, 55 min, 0°C centrifrugation.

(From DeFeudis, 1972f)

## Comment

From the above studies, it has become evident that cerebral morphology changes according to the level of environmental stimulation received by the animal. Similar studies will perhaps reveal that other drugs (e.g., narcotic analgesics) will also be bound more to brain particles prepared from socially deprived animals. Some evidence has already been obtained in support of this notion using $^{3}H$-dihydromorphine (DeFeudis et al., 1977). These studies provide further support for the contentions that changes in cerebral gross and microscopic morphology can be caused by environmental modification, and that the potential for drug addiction may be linked to the environment.

# GENERAL DISCUSSION: ELEMENTS FOR FURTHER EXPERIMENTAL CONTROL

## Species, Strain, and Sex Differences

### *Introduction*

Before entering into a discussion on species, strain, and sex differences, it should be understood that as the phylogenetic scale is ascended, the behavioral patterns of organisms become more complex. This is apparently due to an increased complexity and plasticity of neuronal networks. Social isolation (or individual housing) will therefore be differently interpreted by animals of different species, though the environmental insult, itself, might be quite similar. Even if we assume that the same degree of "isolation" is produced in a rodent and in a human, this insult, which registers as a change in basal "arousal" level (e.g., Welch, 1965; Walsh and Cummins, 1975) and which is modified by the level of dominance of the animal (see Freeman, 1971), may trigger different types of responses. Also, the biological substrates for environmentally induced behavioral changes, though likely to be genetically linked, require environmental modification for their expression. Among the various mammalian species, among various strains within a given species, between the sexes, or among different individuals of a given hierarchical structure, the thresholds for environmental triggering of "abnormal" behaviors vary considerably. With specific regard to the sex of the animal, it has been shown that male mice of various strains become aggressive when subjected to prolonged isolation, whereas female mice do not exhibit this change. The major elements which appear to be related to environmentally induced aggressiveness, or related behaviors, are "arousal" level, body metabolism, and the hierarchical position of the animal.

### *Species Differences*

If modification of the "arousal level" is involved in the behavioral effects caused by social isolation, then all species should exhibit behavioral changes (and other neurobiological changes) when subjected to social isolation. However, if aggressiveness is the only parameter measured, differences will be shown among species. Social isolation of male mice of many strains can cause them to emit aggressive behavior. The male Sprague-Dawley rat (Valzelli, 1967a) does not become aggressive upon prolonged isolation, but the Wistar rat (Bevan et al., 1951) and rabbit (Wolf and von Haxthausen, 1960) do. Social isolation, or a reduction in the mean level of environmental stimulation, induces behavioral abnormalities in many animals (Allee, 1942; Scott, 1947; Yen et al., 1959; Valzelli, 1967a, 1967b; Mason, 1963; Harlow and Harlow, 1962), as well as in man (Leff, 1968; Rosenzweig, 1959; see references in DeFeudis, 1974b, 1974c). The dramatic effects of total social isolation on monkey behavior have been used as models for psychopathological disturbances in humans (see references in Suomi and Harlow, 1972).

*Strain Differences*

Strain differences among rodents exist for many parameters, some of which have been mentioned above for mice and rats. Male mice of various strains exhibit different susceptibilities to isolation-induced aggressiveness and associated cerebral biochemical changes. Although it has been reported that Black CBA and C57 mice do not become aggressive after prolonged isolation (Valzelli, 1967a), the writer has observed this behavior in C57 Black mice (unpublished observations, DeFeudis, 1972; see also Bourgault et al., 1963). However, this strain does not become "as aggressive" upon isolation as the Swiss albino strain.

*Sex Differences*

Female mice do not become aggressive when subjected to prolonged periods of differential housing (e.g., Garattini et al., 1969). Although female mice do not exhibit many of the neurochemical changes displayed by male mice after prolonged isolation (e.g., differences in 5-HT turnover), some other changes do occur in their brains and other organs. It seems necessary to realize that since the "stressor" (individual housing) in these cases has been kept constant between males and females, the difference in behavioral response and in some of the measured neurochemical, or other, parameters may well be due to hormonal influences which modify the animal's interpretation of its environment. Regardless of its etiology, this sex-related difference in response to environment appears to be genuine, and is deserving of further study. There seems to be little doubt that a sex-related difference in response to environmental change exists also in primates (Sackett, 1972).

With more specific regard to neurochemical and pharmacological parameters, Marcucci et al. (1968) showed that prolonged isolation caused a decrease in cerebral N-acetyl-L-aspartate in male, but not in female mice. However, Baumel et al. (1970) showed that both male and female "isolated" mice exhibited an increased resistance to hexobarbital narcosis, and a faster decline in whole body concentration of hexobarbital, even though only the males became aggressive. Welch and Welch (1971) showed that "isolated" females did not fight vigorously when placed together in groups, as did "isolated" males, but that they did examine one another vigorously and run about excitedly. These latter behavioral changes in female mice were accompanied by modest (but non-significant) increases in cerebral NE, DA, and 5-HT levels. Interestingly, these studies of Welch and Welch (1971) revealed that cerebral 5-HT concentrations were higher in female than in male mice of the BALB/c strain, a finding which may be related to the difference in aggressiveness between sexes, since strains of mice with high cerebral 5-HT concentration appear to be less aggressive than those with low brain 5-HT (see NE, DA, and 5-HT section).

## Short- and Long-Term Environmental Changes

Time is one of the most important variables to be considered when one studies the effects that environmental modifications have on brain and behavior. Changes in cerebral chemistry and morphology, as well as in the brain's responses to psycho-active agents, are all heavily dependent upon time. These changes depend both upon the time spent in a previous environment and upon that spent in a novel environment. The effect of time has been clearly demonstrated in studies conducted on the time-course of development of aggressiveness in male mice (see Welch and Welch, 1971). Perhaps the best examples for illustrating the differences between "short-term" and "long-term" environmental changes are the studies on the toxicity of d-amphetamine in mice (see section above). Both short- and long-term environmental changes are important in characterizing the neurochemical and behavioral changes which environment causes in animals. Other temporal processes which are significantly involved in studies of environmental effects on the brain and on behavior are the ages at which animals are weaned, and the age of the initiation of environmental change. In relation to the time variable, one might also consider "chronological" vs "biological" time (or aging). It seems likely that upon long-term exposure to an "impoverished" environment, an animal, due to the effects caused by decreased sensory input (e.g., denervation supersensitivity and cellular agenesis), may possess a brain which is chronologically as old, but functionally younger (i.e., capable of less learning, memory and adaptation) than that of a littermate exposed to environmental "enrichment" for the same period of time. This notion emphasizes the interaction between maturation and environment. It should be further understood that not only can environment produce changes in biological maturation, but also that maturation can alter the organism's (or the brain's) interpretation of a given environment.

## Is There A Morphological Basis For Learning, Memory, and Behavioral Changes?

From the experimental results discussed here, there remains little doubt that a morphological basis for behavioral change exists. Presently, with the advances in technology which have been made and with the accumulation of a vast literature on this subject, the writer is led to suggest that conclusive evidence for the morphological basis of learning, memory, and behavioral change is well within reach. Further research into this area should perhaps be aimed at testing the hypotheses presented herein, and at further validating the experimental results which have been obtained thus far. This research should perhaps be conducted at the primate level.

The answer to the question posed in the title of this section can be derived from the answer to the question of whether or not we believe that the predisposition of an animal (or human) to behavioral disturbance is genetically determined or environmentally determined. It appears that the predisposition, itself, is genetically determined, since certain individuals have high, and others low thresholds to environmental stress. Environmental stress is the trigger and, depending on the threshold of a given individual, a given amount of environmental stress will or will not trigger abnormal behavioral patterns. In most of the studies conducted in the writer's laboratory, the C57 Black strain of mouse was chosen, due to it long history as a control strain, and due to its heavy inbreeding, for the purpose of minimizing effects due to genetic factors, and matched, weanling, littermate, male mice were used in all cases. Nutrition and water supply were identical between "isolated" and "grouped" mice, the only difference being their housing conditions. However, though environment was the only variable altered in the experiments, there seemed to be little doubt that maturation (i.e., *biological maturation*) was altered by differential housing. Evidence for this notion stems from results obtained in subcellular studies. Thus, after the period of differential housing of the mice, the writer believes that biological (or neurochemical) maturation had differed.

Hence, the interaction, though seeming to be most complex between nutritional, maturational, genetic and environmental factors, can be reduced, simply by controlling hereditary and nutritional factors, to an interaction between environment and maturation; maturation is useful for explaining the neurochemical and behavioral effects caused by manipulation of the environment. Further studies, using designs in which genetics and nutrition are controlled, can reduce the interaction to maturational changes which are due to differences in somatosensory input, and which can be observed as changes in synaptic development. These changes are expected to occur as a function of environment, and are presumed by the writer to form the morophological basis of learning, memory and behavioral change in all animals. It is expected that such changes caused by environmental stressors (or differences in socio-environmental conditions) are reversible, but that their reversibility depends upon the length of exposure of the individual to the environmental stressor(s). Hence, cerebral morphology may well undergo a process of continuous change.

# CHAPTER 7
# Sensory Environments and Brain Damage

## INTRODUCTION

Since the sensory environment can modify brain structure, function, and behavior, it might sometimes play a role in the induction of, and recovery from, the effects of brain damage. The evidence for this suggestion will be presented under the following headings: experimental variables, the state of the art, physiological mechanisms, and outstanding questions. The available research data will also be used to speculate upon the etiology of both damaging and repair processes.

## EXPERIMENTAL VARIABLES

### Sensory Environments

Of the many possible types of sensory stimulation, only five appear to have been employed in studies of neural rehabilitation. These five are: light versus dark rearing, social rearing versus isolation, complexity versus isolation, behavioral training versus no training, and active versus passive exploration. Other types of stimulation, such as handling and vestibular stimulation, do not appear to have been used for studies of neural rehabilitation, even though they have proved useful for behavior.

### Brain-Damaging Agents

A major limitation of the field to date is that, with few exceptions, the brain-damaging agents employed prior to studies of neural rehabilitation have been limited to varieties of sensory deprivation (e.g., isolation, and dark rearing). Thus, we have little direct evidence of environmental influences on neural recovery from physical, nutritional, pharmacological, hormonal or metabolic agents, although considerable behavioral recovery may be evident (Walsh and Greenough, 1976b). As will be seen, a number of the changes induced by sensory stimulation are opposite to those produced by several brain-damaging processes, and so heightened stimulation has been the sensory modification most frequently assayed as a therapeutic agent. The emphasis on heightened stimulation in this chapter is not meant to imply that it is always beneficial.

## THE STATE OF THE ART

### Anatomical Recovery

#### *Environmental Complexity*

For the majority of anatomical parameters, it would seem that the effects of prior deprivation can be at least partially reversed by subsequent exposure to heightened stimulation. This has been well demonstrated for cerebral weight, for not only may stimulation enhance recovery, but may also afford protection against subsequent deprivation. This was shown in a series of switchover experiments in which rats were reared from weaning in either isolation or environmental complexity. After various times of exposure, half of the subjects of each group were switched to the opposite environment for an equal period of time, while the remainder continued in the original environment. This procedure resulted in four groups, two of which had continuous exposure to either complexity or isolation, and two of which were exposed first to one and then the other (Cummins and Walsh, 1978). The results from such a 30 day + 30 day design showed that the exposure to complexity was effective in increasing final cerebral weight, whether this exposure was in the first or the second 30 days. Animals that experienced both complexity and isolation (switch-over groups) in either order, showed cerebral weights as large as those of animals that remained in complexity for the full 60 days; and all three of these groups had significantly greater cerebral weights than those of the animals that remained in isolation throughout. Interestingly, this experiment shows not only that complexity was sufficient to overcome the effects of prior deprivation, but also that earlier exposure to complexity afforded protection against subsequent isolation.

While in the above study, both primary and secondary environments exerted a significant effect, Figure 10 shows that in another switchover study, where the animals were reared for prolonged periods (60 days in both primary and secondary environments), the effects of the primary

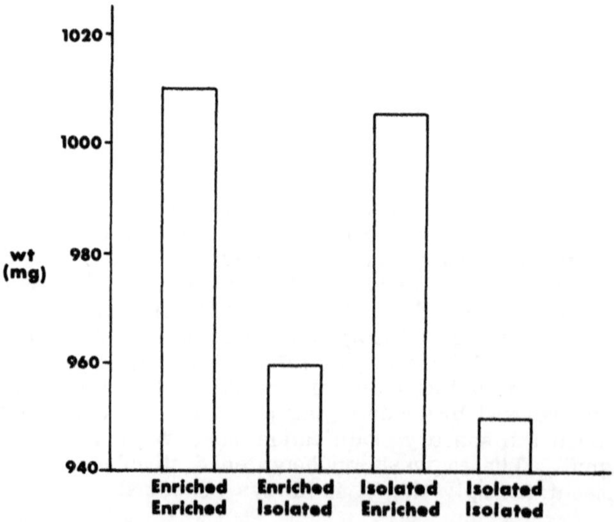


Fig. 1. Effects of 60 days exposure to both primary and secondary environments on cerebral weight. The effect of the secondary environment was significant at the 0.01 level while the primary environment no longer exerted a significant effect, either direct or interactional.

exposure were largely dissipated and the terminal brain weights were determined only by the secondary environment. These studies support some findings of Rosenzweig et al. (1962); 33 days of isolation followed by 48 days of enrichment produced brain weight, and cholinesterase and acetylcholinesterase concentrations, intermediate between those of the littermates who were maintained for the full 81 days in either complexity or isolation. A later study (Rosenzweig et al., 1967) suggested that the effects of 80 days of complexity or isolation were largely removed by a subsequent 50-day period in the opposite environment, so that the terminal values appeared to be determined largely by the secondary environment.

Twenty-one days of the Hebb-Williams maze testing was also found to be effective in overcoming isolation-induced cerebral weight and size reduction (Cummins et al. 1973; Fig. 2). In this study rats were reared under either complexity or isolation for 535 days, and then half of each group was maze tested. For isolates, the testing raised the cerebral values almost to the level of the complexity animals, whereas the complexity reared subjects showed no significant effects of testing. This study suggests that at least some of the neural effects of deprivation may be maintained for considerable periods, yet may still be reversed with relatively short periods of augmented sensory input, a finding with obvious therapeutic implications.

In line with the above studies, showing a signficant but partially transient protection against isolation effects by prior exposure to complexity, are some studies by Bennett et al. (1974). They initially placed weaning rats in either complexity or isolation for either 30 or 80 days of differential rearing. They transferred some animals from complexity to isolation and kept them there for periods ranging from seven to 47 days, in order to measure the persistence of the complexity effects. The effects of 30 days of differential experience appeared to be less persistent than those of 80 days, although it must be noted that the experiments confound the duration of initial rearing and the age at which the subjects were placed in the second environment. After an initial 30 days differential rearing period, cortical weight showed a reduction to about 60% of the original complexity value within seven days, and to about 30% within 14 days, while acetylcholinesterase activity showed a somewhat greater persistence. After an initial 80-day differential rearing period both anatomical and chemical measures once again exhibited asymptotically declining values, but the persistence was greater and remained detectable

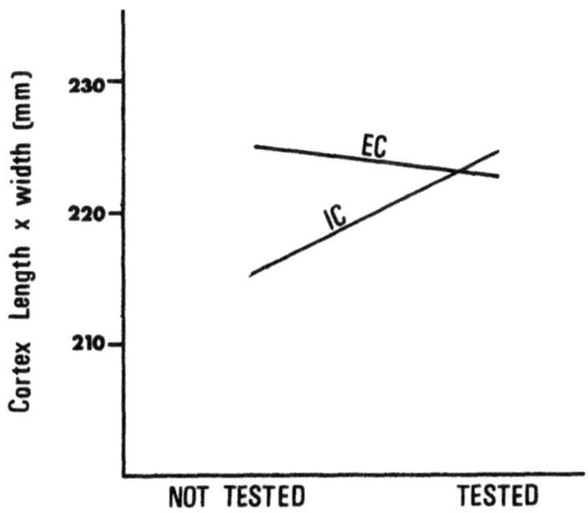


Fig. 2. Testing × environment interaction on the derived parameter of cerebral length × cerebral width. The environment main effect was significant ($p<0.05$) due to the fact that testing exerted an effect on the isolated (4.7%, $p<0.5$) but not on complexity-exposed subjects.

in the occipital cortex after 47 days. Cortical weight persistence = 25%, NS, acetylcholinesterase activity persistence = 65%, ($p < 0.05$). In contradistinction to these trends, cholinesterase activity, which is usually, though variably, reduced by isolation, actually showed an increase after transfer from complexity to isolation. Thus, with the exception of cholinesterase, the anatomical and chemical effects of complexity tend to dissipate when subjects are transferred to isolation, but the degree of dissipation depends on the parameter and brain region under investigation, as well as on the duration of prior treatment and possibly also the age of the subject.

Taken together, these studies suggest that when an animal is removed from one sensory environment and placed in another, the cerebral response to the secondary environment is dependent upon the residual response to the first. Where the period of secondary exposure is short (30 days or less), the subjects coming from the condition of stimulus deprivation attain cerebral weight values close to those of subjects exposed to continuous stimulation. Subjects removed from complexity, on the other hand, show comparatively little loss during the subsequent deprivation. The basis for this differential responsivity has not been established, but one possible explanation is provided by the differential arousability of these subjects. Isolates have been behaviorally observed to be more easily aroused than complexity-reared subjects (Konrad and Melzack, 1975; Cummins et al., 1974; Walsh and Cummins, 1975, 1976a, 1976b), and the physiological arousal response has been hypothesized as one possible mechanism by which sensory stimulation elicits brain changes (see Chapter 8).

Further evidence for the therapeutic potential of environmental complexity has come from the study of recovery after experimental brain damage. There is now considerable evidence that post-lesion experience may exert a therapeutic behavioral effect on a range of species from rats to primates, though the range of behaviors which may benefit (the specificity of the recovery effect) is a more complicated question (Greenough et al., 1976; Goldman, 1976; Goldman and Mendelson, 1977; LeVere and Davis, 1977). Neural recovery is less well demonstrated, but may partially occur, as shown by Will et al. (1976, 1977; Will and Rosenzweig, 1976), who performed bilateral occipital cortex lesions on rats, either at birth or 30 days of age. These animals were then reared for various times in either isolation or environmental complexity. This differential rearing was followed by a period of maze testing, after which the rats were killed and their brains examined for evidence of neurological recovery. The behavioral results reported by these studies are discussed in detail by Will et al. and Greenough et al. (1976).

The rats lesioned at birth were differentially reared either from day five for 60 days, or from day 25 for 40 days, and after approximately 30 days of behavioral testing, the cerebral length and width were measured. It was found that while lesioning reduced both cortical length and width, environmental complexity tended to increase these cerebral dimensions. As with previous reports (Walsh et al., 1971, 1973a), the increase attributable to environmental complexity was greater for length than for width. Especially interesting from a therapeutic standpoint was that the environmental effect was significantly stronger in the lesioned than in the nonlesioned subjects.

When the animals were lesioned at 30 days of age, the effect of environmental complexity was not necessarily greatest in the lesioned subjects. Thus, while environmental complexity tended to increase both the weight and the RNA:DNA ratio of the cortex, the weight increase was the greatest in the sham-operated controls, while the increased RNA:DNA ratio was greatest in the previously lesioned animals. This indicates a degree of variability in the interaction between environmental therapy and brain recovery. Thus, while the overall evidence clearly indicates that environmental stimulation can aid neurological recovery from non-environmentally induced deficits, it remains unclear whether this occurs by a direct effect on the recovery process itself.

*Light Exposure Following Dark Rearing*

Dark rearing evokes a number of anatomical responses analogous to those induced by isolation, but as yet there are few studies of recovery. Reduced cell size in the lateral geniculate and occipital cortex has often been reported (e.g. Gyllensten et al., 1966; Guillery, 1972). Kalil (1973) examined cats which had been dark reared from birth for four to eight weeks, and then light exposed for at least three months. Depth perception, visual avoidance, following, and placing behaviors appeared with only one week of normal vision, and by the end of three months geniculate cell area had normalized. A decrease in the number of pyramidal cell dendritic spines has been found after dark rearing, although this effect was partially transitory. Thus, in animals dark reared from birth there was a tendency to normalization such that by 50 days only a small, though significant, difference was observed, and this normalization was hastened by exposure to light (Valverde, 1967; Ruiz-Marcos and Valverde, 1970).

A number of studies have demonstrated synaptic alterations following dark rearing (Cragg, 1967, 1974; Fifkova, 1970a, 1970b), but the extent of possible recovery is less clear. Vrensen and DeGroot (1974) reared three groups of rabbits: a control group raised under normal lighting, a deprived group raised in darkness for seven months from birth, and a darkness-light group also reared in complete darkness for seven months, but thereafter in normal lighting conditions for 12 months. They measured cortical synaptic length, surface area, number of synaptic contact zones, and the number of synaptic vesicles per terminal. Only the latter showed effects of deprivation: dark reared animals had fewer synaptic vesicles per terminal than normally reared subjects, and this deficit persisted in spite of 12 months of subsequent light rearing. However, not even synaptic vesicles appeared to be affected in the superior colliculis (Vrensen and DeGroot, 1977).

This contrasts markedly with certain findings of Cragg (1967, 1974). He reared rats in darkness until weaning, when half were placed under diurnal lighting conditions for periods varying from three hours to 10 weeks. Light exposure resulted in increased synaptic dimensions in the superficial half of the cortex, with a reversed effect on the lower half. Synaptic numbers per unit area were increased in the lower cortex, but not significantly affected superficially. This was interpreted as due to light-induced synaptogenesis in the deeper layers, and synaptic enlargement superficially in response to light stimulation. Periods of light exposure as short as three hours were sufficient to elicit detectable effects, after a further 48 hours of subsequent darkness. By subjective examination, Cragg (1967) could not identify any qualitative effects on synaptic profiles, vesicles, or regions of apposition. However, in view of the smallness of the anticipated effects, it is not surprising that non-quantitative methods were inadequate to detect them, if indeed they exist.

Cragg's findings suggest a considerably greater synaptic response to light after dark rearing than do the findings of Vrensen and DeGroot. This could reflect a number of factors, especially species differences, differences in duration of exposure, and subject age. Unfortunately, Cragg's studies do not include a group reared continuously in normal diurnal lighting, so that it is impossible to know to what extent the changes he reports represent recovery towards the norm. In any event, it is apparent that cerebral dendritic spines and synapses of light-deprived subjects are susceptible to subsequent light exposure, and that at least for spines, the changes result in a trend toward normalization.

## Chemical Recovery

Chemical studies, though few, tend to support the concept of the reversibility of prior experimental deprivation. The effects of environmental complexity on RNA:DNA ratios in lesioned rat brains (Will et al., 1976, 1977) have already been discussed. Diaz et al. (1978) and Ellison and Bryan (1976) found that rats lesioned by intraventricular administration of 6-hydroxy-dopamine displayed the typical "supersensitivity" effects of hypersensitivity when placed in isolation postoperatively, but not if placed in EC. However, studies of concomitant norepinephrine uptake presented a less clear picture (Diaz et al., 1975, 1978). Meaures of uptake and histofluoresence suggested that enrichment subjects showed the usual complexity effect of greater norepinephrine uptake into cortex, but were also more susceptible to the effects of the toxin, and recovered less fully. This dissociation between these behavioral and chemical measures suggests that brain systems other than norepinephrine are responsible for the enhanced behavioral recovery.

Investigation of protein precursor uptake have been carried out in rats reared from birth to 30 days of age in darkness, and then exposed to light for varying periods (Rose, 1967; Rose and Sinha, 1974). The incorporation of 3-H-lysine into occipital cortex tissue was initially abnormally great, but this persisted for less than 99 hours.

Cholinesterase and acetylcholinesterase activities, in rats reared under complexity or isolation, showed partial or complete dissipation of effects due to the primary environment when the animals were switched to the opposite rearing conditions (Rosenzweig et al., 1967). Similarly, the effects of isolation on these enzymes are partially dissipated by subsequent problem solving tasks (Rosenzweig et al., 1972), a finding reminiscent of that for cerebral weight in the previously mentioned study by Cummins et al. (1973). Geller (1971) reared rats in either complexity or isolation for 30 days, and then rehoused them in pairs for a further 30. Although significant differences in dopamine and norepinephrine concentrations remained, they were smaller than the differences after the original 30 days exposure to differential environments, suggesting that these substances also show at least partial reversibility.

## Electrophysiological Recovery

The electrophysiology of recovery from sensory deprivation is better known than either the anatomy or chemistry. Unilateral and bilateral visual deprivation produce very different effects, and so will be considered separately.

Dark rearing, a form of bilateral visual deprivation, tends to retard maturation of visual evoked potentials. Evoked portentials in rabbits dark reared from birth exhibited a simple form like those of neonatal animals, were rapidly fatiguable (Fourment and Scherrer, 1961; Scherrer and Fourment, 1964), and were superimposed on a low level of background electrical activity (Shilyagina, 1973). Return to normal lighting resulted in a gradual recovery of both the background activity and evoked potentials towards patterns more characteristic of normal adults (Shilyagina, 1973). Meisami and Timiras (1971) have used the electro-shock seizure threshold (the minimum current capable of eliciting evidence of clonic convulsion) as a measure of the development of excitatory-inhibitory balance under differential light exposure. Both continuous lighting and darkness delayed development in the young rat, and both ultimately resulted in a lowered threshold. However, dark but not light reared subjects showed a tendency to normalization, beginning about day 34, and this was hastened by exposure to normal diurnal lighting.

Even a year of deprivation may be partially reversible if it is bilateral. Cynader et al. (1976) reared cats in total darkness or intermittent strobe lighting for a year, followed by 6 - 12 months recovery time in normal lighting. Orientation selective responses barely occurred immediately after deprivation, in about half the occipital cortex units of the recovery subjects, and in most units of controls. While most units in normally reared cats exhibit directional selectivity, this was rarely observed in recovery animals. A general principle seems to be: Deprivation reduces, and recovery enchances, the degree of cortical response specificity.

The interneuronal connections of the visual system are directed by innate mechanisms. However, these connections are highly vulnerable to stimulus deprivation in early postnatal life, especially since retinofugal fibers compete with their homologous contralateral counterparts for innervation of higher order neurons, within the lateral geniculate and the occipital cortex. The magnitude of the competition varies across species, but in the cat, in which it is most evident, early unilateral deprivation upsets it so that the deprived pathways not only suffer a form of disuse atrophy, but also appear to be suppressed by the non-deprived competing pathways such that recovery is minimal (Sherman and Sanderson, 1972; Von Noorden et al., 1976). Indeed, visual cortex lesions may even reduce the perceptual deficit, presumably due to removal of active suppression (Zablocka et al., 1976), and possibly allowing reorganization of remaining tissue (Schmidt, 1976).

The electrophysiological effects of monocular deprivation are detectable at several levels of the visual system. They are less marked in the lateral geniculate than in the occipital cortex, in spite of the fact that the geniculate exhibits greater histological evidence of atrophy. Thus, three months of unilateral deprivation from birth in the kitten did not preclude normal concentric receptive field organization in affected geniculate layers, even though the deprived neurons were significantly smaller. However, some fields had abnormally large centers and contained a reduced number of active cells, as well as some cells which reacted sluggishly (Wiesel and Hubel, 1963).

In the occipital cortex the changes in unit activity were dramatic (Wiesel and Hubel, 1963a; Ganz et al., 1968; Hubel and Wiesel, 1970; Wilson and Sherman, 1975). Some neurons ceased to respond altogether. In the normal cat, about 80% of the striate cells can be driven by both eyes (Hubel and Wiesel, 1962), but this percentage is greatly reduced by unilateral deprivation, and those few cells which remain binocularly driven are dominated by the non-deprived eye. Neurons dominated by the same eye tend to be arranged in ocular dominance columns; following unilateral deprivation, the columns of the deprived eye shrank in size, while those of the non-deprived eye expanded to fill a larger proportion of the cortex. In cells which the deprived eye could still trigger, the response tended to be more sluggish and to fatigue more rapidly. These cells also failed to display the normal response selectivity to oriented line stimuli, that is, they were circular in receptive field and unselective to form or motion. The fact that these effects were less marked with bilateral eye closure (Wiesel and Hubel, 1965) was one of the first suggestions of the possibility of competition between axon terminals from the two sides for control of the postsynaptic cell.

Typically, monocular deprivation requires a long latency period before effects can be convincingly demonstrated (Pettigrew et al., 1973). However, chemical blockage of tonic optic nerve discharge by anesthetics resulted in binocular competition effects within minutes, suggesting that tonic retinal discharge may play an important role in maintaining balance (Kasamatsu, 1976). It also suggest that anesthetic application to the undeprived eye of a formerly deprived animal might have therapeitic potential.

The degree of physiological and behavioral recovery depends on the animal's age and duration of deprivation, and the method of treatment.

If deprivation is maintained throughout the entire sensitive period, which lasts from birth to about three months in the cat, then simply opening the deprived eye has little effect. However, with shorter periods of deprivation, varying degrees of recovery are possible. This recovery is enhanced by closure of the control eye at the same time as the occluded one is opened.

Within the cat lateral geniculate, almost no electrophysiological recovery occurs; however, eye closure reversal elicits a degree of histological rehabilitation, provided that this is done within the first six weeks of life, and enucleation of the opposite eye has similar effects even after this period (Garey and Dursteler, 1975). Within the occipital cortex, reversal of the électrophysiological deficits is more marked provided, of course, that the initial period of deprivation is considerably less than the sensitive period. It may even occur when the animals are exposed to bilateral exposure during recovery, instead of to reverse eye closure (Mitchele et al., 1977). This recovery is most evident as a decrease in the number of unresponsive cells, although the number of binocularly driven cells shows only a minimal change. The number of monocular cells driven by the formerly deprived eye increases, and as it does so, there are concomitant changes in the size of the corresponding ocular dominance columns (Movshon and Blakemore, 1975). Whether this recovery represents simply the activation of formerly ineffective synapses, as has been noted in other regions of the CNS (Wall and Egger, 1971; Merrill and Wall, 1972), or an actual formation of new conncections, is unknown, but it seems reasonable to presume that both may occur.

Periods of deprivation, extending throughout the total duration of the sensitive period, nearly preclude any significant degree of physiological recovery. This suggests that once unilateral innervation of the visual cortex neurons is effected, competition for the postsynaptic cells greatly reduces the chance of functional recovery of the deprived pathways. However, while there was little or no physiological recovery of animals deprived throughout the sensitive period, reversal of eye closure was found to aid behavioral rehabilitation (Dews and Wiesel, 1970; Chow and Stewart, 1972). This behavioral recovery, in the absence of neurophyiological correlates, may seem surprising, but, does appear to be based on the learning of alternative cues for the solving of perceptual problems - an example of behavioral substitution. For example, visually deprived cats have been found to move their heads along the contours of stimuli, thus translating the configurational cues of a discrimination task into a combination of visual flux, visual reafference, and proprioception (Rizzolatti and Tradari, 1971; Ganz and Haffner, 1974). Discriminations such as these are more precarious, and easily disrupted by any change in the sensory environment, thus considerably reducing the generality, transferability and behavioral reserve of the animal. Some electrophysiological recovery can occur in the deprived optic system, as measured by visual cortex unit activity, even well after the critical period, if the non-deprived eye is removed (Kratz et al., 1976). The effect is rapid (within hours), and does not appear to increase with further time, up to a year. The results suggest that functional connections from the lateral geniculate to visual cortex remain after monocular deprivation, but that these are continuously suppressed by the non-deprived system.

Auditory-evoked potentials have been used to investigate the effects of deprivation and subsequent recovery within the auditory system. Rats reared for the first eight weeks of life in a sound attentuated environment exhibited raised auditory thresholds. However, exposure to normal ambient sound resulted in partial recovery within 48 hours, and complete recovery within three weeks (Batkin et al., 1970). The extent of electrophysiological recovery from long-term sensory deprivation in humans is unknown, but it is of more than academic interest that prisoners, expecially those subjected to solitary confinement, exhibit slowing of the EEG (Gendreau et al., 1972.)

Interestingly, different types of deprivation may summate. Thus, cats exposed to monocular deprivation and directional deprivation (exposed to stimuli moving in only one direction), both of which were subsequently reversed, showed summation effects. Recovery from each type depended largely on when, during the respective critical periods of each type of deprivation, the deprivation was terminated (Dewes et al., 1978).

## SUMMARY

In general, the effects of bilateral sensory deprivation can be at least partially reversed by provision of the missing stimuli. For gross anatomical parameters in rodents, there does not seem to be much in the way of a critical period limiting the reversibility of environmentally induced brain changes. However, there has been little investigation of more discrete variables, and it remains possible that the recovery of the size of pre-existing structures, such as increased nuclear and cell body dimensions, may occur more readily than the formation of new cell constitutents (e.g., new dendritic branches and synapses). The brains of deprived animals show a more rapid response to stimulation than do non-deprived subjects, but this effect is transient. As yet there has been almost no success in finding neurobehavioral correlations in these types of recovery studies, and some researchers have called the worth of neural studies into question because of this. This seems premature, since the complexity of the brain is such that one can rarely expect recovery of even a relatively simple behavior to correlate with any one brain parameter.

These comments have referred to recovery from bilateral deprivation. Recovery from the effects of unilateral deprivation seems more difficult largely because of competitive functional suppression. Cells within the retino-cortical system compete with their contralateral homologues for post-synaptic cells, and once the competitive balance has been upset, it is extremely difficult to reverse. Physiological recovery of the deprived eye and its projections is therefore extremely limited, even though there may be considerable apparent behavioral recovery. This physiological-behavioral dissociation is due to the ability to compensate for lost function by using alternative cues.

Future research efforts will undoubtedly aim at enhancing recovery. Several lines of attack seem possible. Closure of the control eye aids recovery, and it may be that this effect can be potentiated. Long acting local anesthetics applied to the dominant visual pathways might reduce competitive inhibition, while application to corticofugal systems, which are partially inhibitory (Sherman, 1974), perhaps at the level of the visual cortex, might be helpful. Anticholinesterase agents and anti-inhibitors such as strychnine could also prove useful. Active exploration and visual reafference have proved helpful in retention of pattern discrimination in rats subjected to two-stage lesions of the visual cortex (Dru et al., 1975), and might prove helpful in visually deprived animals (Greenough et al., 1976).

The monocular deprivation paradigm may prove to be a useful experimental model for a number of significant clinical problems. It has been suggested that humans develop the capacity for speech in both cerebral hemispheres, but that one is subsequently suppressed (Milner, 1973; Geschwind, 1974). The variable degree of recovery from aphasia which follows damage to the dominant hemisphere would thus reflect the varying degrees of contralateral suppression. Similarly, while the limbs are normally activated by ipsilateral corticospinal pathways, they can sometimes be usefully innervated by contralateral tracts when the ipsilateral ones are damaged (Nathan and Smith, 1973) due once again, perhaps, to the release of competitive inhibition. If this is so, the therapeutic task in these cases then becomes one of maximizing the recovery from competitive inhibition, a task for which the monocular deprivation paradigm may prove an excellent model. In any event, because its anatomy and phyiology are so well known, and because the behavioral effects of pathology can be detected with precise perceptual parameters, the visual system will undoubtedly continue to be an excellent model for the study of environmentally induced damage and recovery. Finally, the studies by Will et al., (1976, 1977) demonstrating environment-lesion interactions, present an excellent model for studies of rehabilitation following nonsensorially induced brain damage.

## MECHANISMS MEDIATING RECOVERY

To date, there has been little speculation, and even less empirical evidence, on the mechanisms which mediate environmentally induced recovery. This should come as little surprise, since the mechanisms involved in the induction of environmental effects on normal brains are very much in question. This section examines the mechanisms thought to mediate both environmentally induced brain changes, and recovery after damage. It will also introduce the concept of a function limiting factor, consider interactions between sensory and nonsensory deficiencies, and examine environmentally induced recovery from a reinforcement prespective.

Conceptualization of the mechanisms mediating recovery has traditionally emphasized the physiological approach, although in fact the transduction of sensory stimulation into neural changes probably involves many mechanisms operating at a variety of levels, including behavioral, physiological, and chemical, interacting in a complex interdependent manner. The physiological mechanisms classically thought to underlie recovery are four in number (Luria et al., 1969; Greenough et al., 1976).

1. Functional recovery of damaged neurons
2. Functional takeover by adjacent or contralateral analogous structures
3. Reorganization of neural functional systems
4. Disinhibition of adjacent functionally inactivated neurons

Functional takeover by undamaged tissue is a much debated mechanism which assumes the possibility of latent vicarious function. This concept is closely related to that of redundancy, which assumes that some systems so overlap in their properties that they are partially interchangeable. However, it is apparent that systems vary widely in their recovery potential, and this potential appears to be closely related to the diffuseness and overlap of the system's afferent and efferent anatomical projections (Goldberger, 1974). A host of other factors are surely also involved but not yet identified. Functional takeover has long been thought to occur more readily in younger subjects (Geschwind, 1974), though this is debated by some (Isaacson, 1974), and the extent of the takeover and the role of sensory input are still uncertain.

The concept of the functional reorganization of neural tissue is by no means universally accepted; and some workers, such as LeVere (1975), maintain that recovery is the result of the continued normal operation of spared neural mechanisms. However, among those who accept it, the concept appears to have been used in slightly different ways by Western and Russian workers. Both imply a permanent change in function of the undamaged responsible parts of the nervous system, rather than the expression of latent function. However, recent Western thinking has emphasized factors such as collateral sprouting and denervation supersensitivity (Goldberger, 1974), whereas the Russians have continued to emphasize variations in the nature and sequencing of functional components in the restoration of complex functions (Luria et al., 1969). The latter has been much used therapeutically. Here sensory input of a particular kind is essential: instructional and feedback information from a therapist.

Russian physiologists have made much of the concept of disinhibition in recovery. Following damage, adjacent and contralateral tissue is often found to be functionally depressed, due to so-called protective inhibition, and a decreased ratio of acetycholine to acetycholinesterase in these areas has been postulated (Gazzaniga, 1974). In accordance with this suggestion is the finding that anticholinesterase agents are sometimes therapeutically useful. Functional restoration, for example, of motor activity or speech may occur following anticholinesterase administration, even in cases where the duration of the disorder is measured in years (Luria et at., 1969). This therapeutic effect develops rapidly, is often detected immediately after the first administration, and may even be permanent after a single injection. Sensory stimulation increases the enzymes of the acetylcholine system, and may also increase the acetycholine: acetycholinestrase ratio (Rosenzweig et al., 1962). It might therefore be expected that appropriate sensory input would also be effective in causing disinhibition, and that sensory imput and exercise would produce additive or synergistic effects when combined with anticholinesterases. This has in fact been observed (Luria et al., 1969); for example, in a case where initial exercise therapy was ineffective in motor recovery from brain damage, a single injection of anticholinesterase elicited partial recovery, which thereafter increased with exercise.

The fact that drug-exercise combinations may elicit recovery, where either component alone proves ineffective, suggests the possibility of some kind of therapeutic threshold in which initial disinhibition may require massive stimulation, after which maintenance may be relatively easily obtained. Such a possibility suggests that some types of brain damage may benefit from an initial vigorous multimodality therapeutic assault, rather than from the graduated rehabilitation which is the norm. It also calls into question the therapeutic nihilism which holds that rehabilitation programs can have little effect on speech or motor recovery following brain damage such as stroke (Sarno, 1970; Stern et al., 1971), which has inclined Western neurologists not to attempt therapeutic trails using anticholinesterases.

The arousal response has been advanced as a physiological mechanism mediating both environmentally induced changes and brain damage recovery (Walsh and Cummins, 1975). Reafferent stimuli and stimulus novelty may be particularly important in inducing effects by virtue of their arousal - eliciting nature. There are a number of tantalizing reports of amphetamines facilitating the recovery of discriminative behavior in rats (Braun et al., 1966) as well as both accelerating and making more complete the recovery of motor function in monkeys; while in other monkeys, sedatives such as barbiturates had the opposite effect (Watson and Kennard, 1945).

If the arousal mechanism does represent a common link between environmentally induced changes and recovery from brain damage, or if arousal mediates the production of environmentally induced changes which compensate for brain damage, then the study of arousal-induced changes in normal brains should be relevant to the therapeutic situation. For example, it is known that the effects of environmental complexity on brain weight and enzymes of the cholinesterase system are enhanced by stimulant drugs, simultaneous presentation of social and nonsocial stimuli, stimulation during the most behaviorally aroused part of the diurnal rhythm (Rosenzweig et al., 1972), and stimulus novelty (Brown and King, 1971). Effects are dependent on direct contact with stimulus objects (Ferchmin et al., 1975), and reafferent stimulation may be particularly important (Walsh and Cummins, 1975). Indeed, active, as opposed to passive, exploration has been found to be important to recovery from occipital cortex lesions in rats (Hein, 1970; Dru et al., 1975). Similarly, the variety and kinds of objects available to the normal human infant may enhance arousal and cognitive development (Yarrow et al., 1972; Korner and Thoman, 1972; Ainsworth, 1973; Hunt, 1976), though there is evidence to suggest that it is only when they become incorporated into a social intereaction that they exert significant effects (Williams and Scarr, 1971; Clarke-Stewart, 1974). These findings suggest a number of principles which might be used to guide research and clinical practice in the therapy of brain damage. Perhaps, therefore, an optimal therapeutic regime might include direct and reafferent stimulation of a novel or changing nature, given in a social context which provided feedback and rewarded high stimulation levels, at times of the day at which arousal was maximal, perhaps aided by stimulant drugs.

A large number of anatomical changes occur within the brain both during recovery from damage and following sensory stimulation. It is possible to conceptualize these as mediating mechanisms for environmentally facilitated recovery, although at the present time this is largely speculative. For example, aberrant major fiber tracts may develop, following certain types of brain damage in young subjects. In the newborn kitten, monocular enucleation results in the axons from the remaining eye spreading to layers of the lateral geniculate from which they are normally excluded (Kalil, 1973). In the newborn rat, unilateral damage to the corticospinal tract results in the hypertrophy of an uncrossed corticospinal pathway (Hicks and D'Amato, 1970; Castro, 1975). It is possible that this might be the neural basis for the finding that animals so damaged at birth are sometimes capable of greater recovery than ones damaged later in life. The functional significance of these aberrant fiber tracts is unknown. Dejerine (1914) reported a similar phenomenon in humans with infantile hemiplegia, and it seems quite feasible to expect that sensory stimulation might be capable of modifying the extent of their functional development.

A second possible anatomical mechanism is that of axonal sprouting. Typically, this has been considered a regenerative response of proliferation and migration of axon collaterals into areas deprived of their normal input, although it is possible that the brain undergoes ultrastructural remodeling throughout life (Sotelo and Palay, 1971). The axonal sprouting may occur from transsected neurons or as collateral sprouting from intact cells, although only the former has been examined in detail in the CNS (for reviews, see Guth, 1974; Moore, 1974). The implicit assumption has often been that such sprouting aids recovery by replacing lost input. However, the value of this replacement, which may even be from another system entirely, is uncertain. Indeed, preliminary studies by Isaacson (1974) suggested that chemical inhibition of sprouting actually ameliorated some of the usual consequences of brain damage.

The possibility of future pharmacological interventions to therapeutically modify this sprouting seems a very real possibility. However, in view of Isaacson's findings, it is uncertain whether the optimal modification will entail increasing or decreasing sprouting, or both. Conceivably, stereotactic injection might allow a highly selective, regionally specific modification of the sprouting process, that will ensure regenerative input from one particular system in those areas where regeneration from inappropriate systems is likely to occur, for example, the hippocampus (Lynch et al., 1974, 1975). Agents that have been found to enhance axonal regeneration include triiodothyronine and nerve growth factor, while concanavlin A (a lectin that binds to glycoproteins) appeared to reduce adjacent micro-cavitation (Bjorklund and Stenevi, 1972; Fertig et al., 1971). The dynamic nature of the remodeling of the damaged nervous system is suggested by studies of Bernstein (e.g., Bernstein and Bernstein, 1973), who found that the number of boutons terminaux on spinal cord neurons below the level of a lesion, declined for the first 20 postoperative days, recovered for the next 10, and then exhibited a secondary decline for a further 30 days. This suggests the possibility of a critical, or at least sensitive, period for optimal therapeutic intervention.

Although sprouting typically has been considered a regenerative phenomenon, it seems possible that it may occur continuously throughout life as a consequence of an ultrastructural dynamic equilibrium with which the brain responds to functional demands. Consequently, it may also occur following sensory stimulation, which can certainly increase dendritic and synaptic numbers. It seems reasonable therefore, to suggest that sensory environments may modify axonal sprouting in both damaged and undamaged neurons, and that this may be one of the mechanisms mediating environmentally facilitated recovery.

A number of authors have reported modification of dendritic length, arborization, spatial orientation, and dendritic spines following sensory stimulation and deprivation (Globus et al., 1973; Greenough et al., 1973). Valverde (1967) found that the spatial orientation of visual cortex dendrites following enucleation of the eye was altered in such a way as to reduce the number of dendrites in the functionally deprived layers, and increase them in others. It seems possible that the mechanisms of dendritic and axonal reorganization might form a basis for functional takeover by perifocal tissue. Whether the effectiveness of this takeover might be enhanced by sensory stimulation remains to be determined.

One further possible anatomical mechanism remains, but must be regarded as highly speculative at the present time. This is the postnatal formation of new neurons. Altman et al. (1968b) have demonstrated that at least in the rodent, neuron formation continues after birth in the anterior cerebrum and dentate gyrus. Both these areas have proved morphologically sensitive to environmental complexity (Walsh et al., 1973a), and it seems possible that this may reflect an environmental modification of the neurogenesis. If so, then modification of this process might also occur in environmentally facilitated rehabilitation, at least in younger subjects.

## The Function-Limiting Factor

The basic tenet of this discussion has been that the nature and amount of sensory stimulation may set a limit to the function and development of the brain. Obviously this is only one example of a large range of other factors (genetic, nutritional, hormonal), any one of which may determine the limits of functioning in the normal brain or limit recovery after brain damage. In other words, the functional ceiling will be determined by the most deficient factor; what we might call "The Function-Limiting Factor." Orthomolecular psychiatry (Pauling, 1968), with its emphasis on providing supplements of assumed function-limiting molecules, may be seen as one derivative of a function-limiting factor concept, though the latter includes the nonmolecular factors of sensory stimuli. Since diminished neuronal metabolic activity would be a common end point for several types of deficiencies, this function-limiting factor hypothesis helps to explain certain similarities in the brains of animals suffering from diverse types of deficiencies. For example, the reduced neuronal size, RNA:DNA ratio, amounts of protein, and less extensive dendritic arborization can be seen as the resultants of a final common pathway metabolic slowdown produced by any one of a number of limiting factors, both sensory and nonsensory.

The concept of a function-limiting factor raises several important therapeutic questions. The first of these concerns the nature of the interaction between two or more limiting factors. Presumably the interaction can be additive or interactive, so that the effects of one factor exacerbate or protect against the effects of another. Where neural insults follow one another sequentially, it is important to know the nature of this interaction, for the preventive purposes of identifying subjects at particular risk following exposure to the first trauma. For example, if infants with a history of perinatal trauma or malnutrition subsequently exhibit reduced tolerance to variations in sensory input, then obviously their environments and upbringing will require specific precautions. Although the data is preliminary, it is interesting to note in light of this dimension, the findings of reduced apneic espisodes and accelerated weight gain and motor development in premature infants given tactile and kinesthetic stimulation (Solkoff et al., 1969; White and LaBarba, 1976; Korner et al., 1975).

The nature of this interaction may be even more important when we come to individuals subject to simultaneous risk factors. If the nature of the interaction is additive, then at times of risk, we must be particularly aware of subjects exposed to two or more risk factors. However, if the nature of the interaction is protective, that is, if the presence of deficiency A protects against the effects of deficiency B, the intervention with factor A prior to remedying deficiency of factor B may be disasterous. Such situations are not uncommon in medicine; as, for example, in the alcoholic undergoing withdrawal who is susceptible to both glucose and thiamine deficiencies. If glucose is administered without thiamine, the increase in metabolic demands may then precipitate life-threatening central nervous system manifestations of thiamine deficiency. It seems quite possible that a similar situation could occur where one of the deficient factors is sensory stimulation. Perhaps stimulation should be kept to a minimum during the acute phases of malnutrition, hypothyrodism, and trauma, in order to minimize metabolic demands on marginally functioning neurons, even though stimulation in the post-acute rehabilitative phase might be helpful. There is some indirect evidence to support this concept. Following brain damage, the disordered system may fail at first to respond to any stimulation; but, during recovery, may evince a paradoxical response known as supramarginal inhibition (Luria et al., 1969). During this phase, weak stimuli elicit adequate responses, whereas normal and strong stimuli evoke no reaction, presumably due to functional overload of the neurons' reduced metabolic capacity.

Another question concerns the nature of the function-limiting factor in an apparently normal or recuperating brain, without evidence of nutritional, hormonal or sensory deficiencies. Presumably if the nature of this deficiency could be determined, then its addition would both hasten and increase the ceiling of development. Mathies et al., (1973) have suggested that the rate-limiting factor for RNA synthesis, which is one of the initial stages in the proposed chemical mechanism mediating the transduction of sensory stimulation into altered brain structure, is pyrimidine precusor availability. A worthwhile experiment would be to determine the effects of precursor administration on recovery from brain damage. It may also be possible to extend the concept of limiting factor interaction to the therapeutic use of anesthetic agents and other brain activity inhibitors during acute brain damage situations. For example, perhaps brief anesthesia during the acute phase of thromboembolic stroke or perinatal anoxia would reduce the metabolic demands of the oxygen-deprived tissue to a level where some additional neurons would survive, until the insulting agent could be removed. This is not to be confused with the approach of Ivanow-Smolenski (1954), who anesthetized war trauma victims after the brain injury in an effort to reduce Pavlovian inhibition, which he felt to be responsible for many behavioral deficits.

## Effective Stimuli and a Reinforcement Perspective

A major question for studies of environmentally facilitated recovery concerns the precise nature of the facilitating stimuli. If this is known, then an orthosensory environment or therapeutic regime can be designed. However, all we have currently are some preliminary data concerning the stimuli which are most potent in eliciting environmental complexity effects in normal brains. It is apparent that subjects must interact physically with the stimuli, and it seems that reafferent stimulation may be particularly important (Ganz, 1975; Walsh and Cummins, 1975; Ferchmin et al., 1975; Dru et al., 1975). The power of reafferent stimulation raises two important points: firstly, that the individual serves largely as his own stimulus source; and, secondly, that feedback information may be especially valuable.

Feedback may be even more important in therapeutic situations. Certainly it is one of the major functions of a therapist during clinical rehabilitation (Luria et al., 1969). A possible important future source of a special type of reafference may be biofeedback which, in preliminary studies, has proved promising for a variety of brain dysfunctions, ranging from the motor to the electroencephalographic. In the presence of a therapist, or of biofeedback, each performance becomes a source of further corrective information and reinforcement aimed specifically at the function under consideration. Ultimately, performance and perceptual sensitivity may be increased to a level at which feedback augmentation by therapist, or instrumentation, are no longer necessary.

If reafferant input is important, and the individual serves largely as his own stimulus source, then the effects of environmental stimulation may be mediated by stimulation from the behavioral changes with which the individual responds to the environment. Thus, in addition to exerting a direct physiological effect on the nervous system, the sensory environment may also elicit recovery via its potential for eliciting therapeutic behavior. This raises the possibility of a whole new conceptual framework within which to view the effects of environmental stimulation - namely, a reinforcement framework. Within such a framework, the therapeutic task becomes one of identifying the nature and magnitude of the behaviors associated with recovery, and designing environments which are optimally reinforcing to them; that is, which tend to sustain or increase the frequency of appropriate behaviors at optimal levels.

The reinforcement framework allows a new perspective on one of the major, and most neglected, therapeutic problems, that of maintenance of therapeutic gains (Thoresen and Mahoney, 1974). Intervention programs to date have emphasized the provision of short term stimulation to the subject. Where this leads to behavior which is intrinsically reinforcing (e.g., walking), then it is likely that these behaviors will generalize to other environments. However, as Hunt (1976) notes, other behaviors, such as seeking intellectual stimulation, may not be reinforced in home environments, a fact which may account for the transient nature of the intellectual gains made in programs such as Head Start. Once again, therefore, the therapeutic task becomes one of identifying therapeutic and therapy-sustaining behavior, and designing optimally reinforcing environments.

## IMPLICATIONS, QUESTIONS AND FUTURE RESEARCH

The study of neural responses to therapeutic environments is obviously at an early stage. Relatively few factors have been investigated as yet, and one of the immediate priorities must be to extend studies across a wider range of variables. The first of these variables is the nature of the lesion. Its pathophysiological form (trauma, endocrine, etc.) and acuity of onset are both important; there is already evidence to suggest that the sensory environment is important in determining the extent of behavioral deficit which follows staged lesions (Isaac, 1964; Worthington and Isaac, 1967; Greenough et al., 1976). The site of the lesion will obviously determine the nature and extent of the behavioral deficit, and presumably the optimal variety of sensory input. In fact, future clinical studies will undoubtedly move in the direction of a more precise neuropsychological assessment of deficits, in order to determine the precise nature of the optimal rehabilitative input.

The genetic, developmental, and experiental history of the individual must also be investigated, since it is already apparent that there are wide individual differences in the behavioral sequelae of brain damage and sensory stimulation. Presumably, similar differences in therapeutic responsiveness will also be identified, and comparison of responders and nonresponders may provide some insight into the therapeutic process. Differential responsiveness according to age remains an unresolved question, but studies such as those of Cummins et al. (1973) indicate a far greater degree of cerebral plasticity in geriatric animals than had previously been thought possible, and argue against the therapeutic nihilism which has so often permeated thinking about geriatric rehabilitation.

The brain regions and parameters under investigation also need extension. While changes within the focal area of damage are important, the Russian emphasis on perifocal protective inhibition and therapeutic disinhibition suggests that investigation of perifocal areas could be fruitful. The effects of rehabilitation on unrelated neural areas have been accorded little importance. However, recent studies have revealed the gradual development of widespread ipsilateral brain atrophy following localized cerebral damage (Isaacson, 1974; Will et al., 1977), or lobotomy (Geschwind, 1974). In the case of lobotomized patients, this atrophy parallels a slow decline in intellectual function (Hamlin, 1970), and it may be that some behavioral sequelae of brain damage thought to be due to changes in the primary focus may actually reflect this type of secondary atrophy. On the other hand, compensatory hypertrophy is a possibility. It has been suggested as a basis for the hypertrophy of the auditory cortex in mice (Gyllensten et al., 1966; Ryugo et al., 1975), for the reduced occipital cortex (Bennet et al., 1964). At the perceptual level, some deprived humans have been found to possess greater visual acuity (Bross and Zubeck, 1975). Finally, of course, it is obvious that we need information on a wider range of anatomical, physiological and chemical parameters, obtained preferably from the same subjects and in conjunction with behavioral studies. As always, it is easier to suggest than to do such studies, but it is hoped that the huge amount of information and the greater significance for meaningful interpretation which such studies afford, plus their potential for clinical application, may prove sufficiently rewarding to seduce some experimenters.

It is important to learn to what extent the findings derived from studies of environmental stimulation on normal brains are applicable to the brain-damaged. For example, if the effects and mechanisms are similar, then optimal treatment programs can be planned rationally on the basis of what is already known concerning normal brains. It also seems possible that some of the deprivation recovery studies already performed on the visual system may serve as useful clinical models. Monocular deprivation may serve as a model for aphasia rehabilitation following damage to the dominant speech center.

Another line of questioning concerns the temporal characteristics of recovery. The question of when to intervene may be very important in view of the preceding discussion on possible interactions between function limiting factors. Similarly, if we accept the hypothesis that initial therapeutic stimulation must exceed a certain threshold level to be effective, then in some situations it might be appropriate to apply simultaneous, multimodality stimulation, rather than a graduated sequential program. Since certain cerebral effects of stimulation are partially transient, the question of maintaining gains arises. Presumably this could involve a number of "booster" sessions (reinstatement), or more effectively, by reinforcement of those behaviors such as stimulus seeking, which are themselves effective in assisting recovery.

Assuming that environmental recovery programs continue to prove effective, one of the futher research foci will surely be how to augment such recovery. One of the possibilities relates to the nature of the stimuli employed. To date, a small range of therapeutic environments have been employed, and a number of other sensory modalities such as vestibular stimulation (Korner and Thoman, 1972) may prove effective in some dysfunctions. Another approach may be to match more precisely the type of stimulus to the neuropsychological deficit. Other approaches will be to try to determine and treat the function-limiting factor for each individual case, while combinations of sensory and pharmacological interventions also may prove useful.

Finally, an effort must be made to educate researchers and clinicians alike of the sensitivity of the nervous system, both normal and damaged, to its sensory environment. It took decades for researchers to begin to appreciate the extent and subtlety of environmental influences on behavior, and even longer to appreciate the plasticity of brain anatomy and chemistry. Similarly, environmental sensory influences on the damaged brain have been almost completely ignored in the past, but we continue to ignore them at our own risk.

# CHAPTER 8

# Mechanisms Mediating the Production of Environmentally Induced Brain Changes

Despite the continuing discovery of changes in brain chemistry and anatomy following exposure to environmental stimulation, no mechanism advanced to date has proved satisfactory in explaining the mediation of these changes. Postulated physiological mechanisms include stress, endocrine, maturational, and learning factors. Each of these mechanisms and their shortcomings have been discussed by Rosenzweig (1969; Rosenzweig et al., 1971) and so are dealt with only briefly here. Evidence is then presented to suggest that the arousal reaction (orienting response) may be a major mediating mechanism by showing that the arousal histories of stimulus-enriched and stimulus-deprived animals are different, and that the properties and consequences of the arousal reaction are such as may account for much current knowledge of environmental effects.

## PROPOSED MECHANISMS OF ENVIRONMENTAL EFFECTS

### Stress

Of the proposed physiological mechanisms, the one which has been advanced most frequently is stress. Isolation has been described as being stressful (Hatch et al., 1963; Valzelli, 1973a, 1973b, 1973c), though the generality of the so-called "isolation stress syndrome" remains a matter of debate (Rosenzweig and Bennett, 1969). On the other hand, it might be argued that information overload and high population density may be stressful for complexity-reared subjects. This raises once again the perennial questions, what is stress? and what stimuli are stressful?

There seems to have been a tendency, particularly marked in the "early handling" literature, to view all experimental manipulation as stressful (Russell, 1971) and to define stress in circular terms. It seems high time to introduce a behavioral definition of stressful stimuli. A simple one might be: those stimuli which act as negative reinforcers. The need for an adequate definition becomes even clearer when we turn to the postulated mechanisms mediating early handling effects. Several experiments have been performed in order to test the stress hypothesis, but unfortunately they tend to reflect the general confusion concerning the nature of stress. Thus, since experimental situations thought to be highly stressful do not elicit a pattern of brain responses similar to those of complexity e.g. footshock (Rosenzweig et al., 1971), tumbling in a revolving drum (Riege and Morimoto, 1970), or a combination of both (Riege and Morimoto cited by Rosenzweig et al., 1972), it has been concluded that stressful stimuli play a minor or nonexistent role in eliciting complexity-deprivation effects.

However, in point of fact, the above situations differ in a number of stimulus dimensions from the complexity-deprivation experiments, and it is of course impossible to determine to what extent the aversive aspects of the situations are effective in eliciting the observed effects. Thus, while stressful stimuli certainly modify brain anatomy and chemistry, there is as yet no evidence to suggest that they are any more effective in this than are nonstressful (non-aversive) stimuli which elicit the same degree of arousal, or that stressful stimuli are implicated more than nonstressful stimuli in the production of complexity-deprivation differences. Therefore, the stress hypothesis of complexity-isolation effects does not appear to be supported by either theory or experimentation.

## Hormones

Hormonal mediation was suggested first by Sperry (1968), but recent studies, which were reviewed in Chapter 5, have suggested only a minor role for the endocrine system. Levitan et al. (1972a, 1972b), largely in an effort to test the stress induction hypothesis, found that corticosterone did not elicit the same pattern of subcellular leucine incorporation as that evoked by the complexity-deprivation paradigm. Rosenzweig et al. (1972) found that hypophysectomy did not prevent the development of environmentally induced differences in brain weight and chemistry, in spite of an overall reduction in brain growth. These results suggest that while hormones may be involved, their role is a small one.

## Maturation

It also seems unlikely that the differences can be related to different rates of central nervous system maturation. Several of the environmental effects are opposite to those which occur during maturation, and both Reige (1971) and Cummins et al. (1973) have shown that the environmental effects can be induced in adult animals.

## Learning

The role of learning processes and memory storage in the production of brain changes remains unknown. One would certainly expect the enriched animals to learn more than their isolated counterparts, and Rosenzweig et al. (1967) have argued that the transient anatomical and biochemical changes may reflect the active processes involved in long-term memory formation. The increased nuclear and perikaryonal size of the enriched animals' neurons would thus reflect the increased metabolism necessary to support these processes, but later would regress, leaving only the correlates of long-term memory to distinguish the differentially reared brains. Ambiguous support is given by studies of the neurological concomitants of learning tasks. Rats given problem-solving tasks, mainly of the visual discrimination type, in Skinner boxes, have been found to develop changes in total cortical AChE and ChE when compared with yoked controls (Rosenzweig et al., 1971). The effects were small and dissimilar to those of enrichment, in that the increase in AChE was as large as that of ChE. Similarly, rats runing 60 trials per day in an automatic maze, or 20 trials per day in a modified Hebb-Williams maze for a month yielded only small and mainly nonsignificant changes (Rosenzweig, 1978), and the addition of daily maze training to the enrichment procedure was without significant effect (Brown and King, 1971; Rosenzweig, 1966). This could be taken to indicate that learning plays only a small role; but, on the other hand, it might be argued that less learning occurs in such formal tasks than following exposure to the wide range of stimulus objects and other animals found in a more complex environment.

However, a far more important consideration is that it is by no means certain that experiments such as these are in fact measuring the biochemical substrates of learning. As Greenough and Maier (1972) have so nicely pointed out in their methodological critique of learning studies, standard control procedures ensure neither that no learning occurs in the control group (Maier et al., 1969), nor that the nonlearning aspects of the procedures are the same for both groups (Weiss, 1968). Thus, all that can really be said of such experiments is that the observed brain changes reflect differences in environmental stimulation, but that the degree to which these changes are mediated by differential learning is unknown. At the present time, one can only conclude that learning is almost certainly involved in the induction of brain changes resulting from environmental stimulation in general, and from environmental complexity in particular, but that the extent of this contribution is almost entirely unknown.

## Summary

Efforts to mimic environmental effects with more specific forms of stimulation have not been successful, suggesting that no single operationally defined process appears to underlie these effects. An alternative which is explored here is that a combination of activation of sensory systems, especially by reafferent stimulation, and nonspecific activation or arousal, underlies some proportion of the differential brain effects.

## EVIDENCE FOR THE AROUSAL HYPOTHESIS

The hypothesis that the arousal reaction or orienting response may be of importance in the production of the complexity-deprivation brain differences was suggested by a number of observations. The addition or changing of toys in the complexity cages each day was found to result in a marked hyperactivity, which was maximal within a few minutes and gradually decreased over the next few hours. During this period of hyperactivity, the complexity subjects engaged in considerable, and at times almost frenzied, exploration of the toys, as well as in considerable social interaction, and were obviously highly aroused. This response continued throughout the experiments, but became less marked and habituated more rapidly with longer duration of differential rearing (Cummins et al., 1972; Ferchmin and Eterovic, 1973). It was also noted that on removal from the home cage at the completion of differential rearing, isolates appeared highly emotional and aroused, by comparison with the placid complexity subjects. This observation was confirmed by open field testing (Cummins et al., 1971, 1972, 1974) and would appear to be a common finding (Fox, 1967; Fuller, 1967; Konrad and Melzack, 1975; Melzack, 1968, 1969; Walsh and Cummins, 1976b). Thus, it would appear that the enriched animals are subject to greater arousal during the course of the differential rearing period, but are subsequently less apt to exhibit arousal reactions on exposure to a novel environment.

The arousal reaction has been reviewed by Lynn (1966), and its major neurological property is a transient increase in the electrical activity of the brain. The importance of this lies in the fact that, as will be seen shortly, it is almost certainly the transformation of neural excitation into biochemical metabolic activity which is important in the induction of cerebral effects. The arousal reaction is elicited most readily by stimuli which are novel, or which hold special significance to the animal, either by virtue of innate biological determinants or of learning. With prolonged exposure to an initially novel stimulus, habituation occurs; and with repeated presentation of the stimulus, the rate of habituation increases and the marked generalized electrical activity of the cortex diminishes, to be replaced by a more localized discrete cortical response to the stimulus.

The relevance of these properties to the induction of enrichment-deprivation brain differences is as follows. In view of the elicitation of the arousal reaction by novel stimuli, it would seem reasonable to suppose that the addition of toys to an otherwise barren laboratory environment would present a number of initially novel stimuli which might be expected to induce arousal responses. Brown and King (1971) emphasized the importance of novelty in the enriched environment by demonstrating that the variety of stimulation in the complex environment was more effective in eliciting changes in brain cholinesterase than was the sheer number of stimuli. The previously mentioned observations of hyperactivity following toy changing also support this idea, and the gradual habituation of this activity both within and between days fits well with the habituation properties of the arousal. In fact, since the animals habituate to the arousal properties of the toys within a few hours, it might be expected that limited periods of daily exposure would be only slightly less effective than exposure throughout the 24 hours. In fact, only two hours suffices (Rosenzweig et al., 1971). Habituation of the arousal reaction also occurs on successive days, with localized cortical electrical activity replacing the initial generalized response (Lynn, 1966). Therefore, it might be expected that the physical brain changes would be at least partially transient, and such has been found to be the case. For while several areas of the cortex show an initial response, it subsequently diminishes in all areas except the occipital. Possibly in the rat this represents the region of localized post-habituation cortical analysis of at least some of the complexity stimuli, and a desirable experiment would be the monitoring of EEG activity during differential rearing to determine the accuracy of this hypothesis.

## Neurostasis

A further factor accounting for the transience of effects may be arousal homeostasis. It is well known that too much or too little arousal is deleterious to a number of central nervous system functions, ranging from cellular metabolism to behavioral performance. The behavioral sphere has received the most attention, and several workers (Hebb, 1955; Malmo, 1959; Schlosberg, 1954) have postulated an inverted U-shaped curve to describe the relation between arousal level and behavioral efficiency. The optimal level may of course vary between individuals, between different neural systems, and from one task to another (Duffy, 1957; Lacey, 1967; Schlosberg, 1954). To date, it seems that the inverted U concept has been limited to behavior, but it may well also be applicable across multiple levels of central nervous system functions. Certainly at the neuron level, too little or too much activity results in reduced cell size and amounts of various constituents, including protein and RNA (Geinismann, 1971; Geinismann et al., 1971; Hyden, 1943).

In view of these facts, one might anticipate that the brain would employ a variety of homeostatic mechanisms designed to maintain this intermediate optimal arousal level. This concept of arousal homestasis is far from new, originally surfacing in the nineteenth century, particularly among the physiological school of Helmholtz, as the "law of constancy". "The nervous system endeavors to keep constant something in its functional condition that may be described as the sum of excitation" (Breuer and Freud, 1892/1955).

Indeed, in the last century this concept formed the basis of Freud's subsequent attempts to design a "scientific psychology" (Pribram and Gill, 1976). Under this general constancy law, or "neurostasis," as it might be called, one can perhaps subsume more specific laws and hypotheses, such as Cannon and Rosenbleuth's (1949) law of denervation hypersensitivity, the concept of stimulation deprivation hypersensitivity (Konrad and Melzack, 1975; Prescott, 1970a, 1970b), Sharpless' (1970) model of drug dependence, and Levine's (1969a, 1969b) stress immunization hypothesis.

Arousal homeostasis may be mediated at three levels: behavioral, neurological, and endocrine. A number of researchers have noted that organisms seek or avoid stimulation in an effort to obtain an optimal level of arousal (Berlyne, 1960; Hebb and Thompson, 1954; Leuba, 1955; Ludwig and Stark, 1973). Others (McClelland et al., 1953; Young, 1959; Pfaffman, 1960) have stated similar theories in hedonic terms. In addition to this control of stimulus level by behavior, the arousal effects of the stimuli may also be modified by the central nervous system directly. Examples include habituation to the effects of both sedative and stimulant drugs, such as the barbiturates and amphetamines. In part, this reflects hepatic enzyme induction and hence increased detoxification, but a further component is centrally mediated (Baumel et al., 1969). Similarly, electroconvulsive shock threshold rises with repetition (Pryor et al., 1967), and even the isolated cerebral cortex exhibits this type of neurostasis (Grafstein and Sastry, 1957; Sharpless and Halpern, 1962).

Evidence of arousal homeostasis compensating for altered levels of sensory input has also been obtained. At the synaptic level, norepinephrine uptake into cortical synaptosomes is altered by fighting in such a manner as to result in arousal homeostasis (Hendley et al., 1973; Welch et al., 1974b). Electrophysiologically, a number of animal studies (reviewed by Zubeck, 1969) have demonstrated a slowing of cortical EEG rhythms, similar to those observed in sensorially deprived humans, following visual, auditory, or olfactory deafferentation. An interesting study by Beteleva and Movikova (1961), however, demonstrated that while olfactory deafferentation depressed EEG rhythms, there was an increase in reticular excitability which normalized over the next 1-2 months.

This neural homeostasis may also be evidenced by endocrine function, since at least some hormones (the best known being those of the adrenal) reflect levels of both past and present sensory stimulation and arousal (Michael and Gibbons, 1963; Vogt, 1960; Zuckerman, 1969), and in turn feedback to modulate central nervous system arousal. The overall effect of these three levels of control is neurostasis.

In view of this neurostasis, one might anticipate that an environment rich in arousal-inducing stimuli would eventually result in a reduced reactivity to these and similar stimuli. At a psychological level, this might be thought of as an example of stimulus generalization. It might also be conceptualized within the framework of Sokolov's (1960) stimulus-matching neural model, according to which each stimulus elicits a characteristic neural model of itself. Restimulation results in the activation of a process which attempts to match the stimulus effects with the prior models. Matching results in recognition, while failure to match elicits an orienting response. Complexity-exposed subjects, with their history of greater prior stimulation, would be expected to develop large numbers of such models, and hence to be less susceptible to an arousal response than would isolates. On the other hand, a stimulus-deprived environment would be expected to result eventually in sensitization, so that orienting responses would be more easily elicited. The net effect of these alterations on the arousal threshold would be that the arousal potentials of both types of living environment would become more similar for their respective inhabitants, the longer the duration of exposure. This factor would result in a further mechanism, in addition to simple habituation, accounting for the transient nature of the observed brain differences.

This reasoning is supported by behavior following removal from the home cage, at which time the formerly stimulus-deprived isolates, by comparison with the enriched subjects, exhibit marked hyperarousal. It might be expected that this heightened arousability would mean that brain changes would be more easily elicited by sensory stimulation in former isolates than in other subjects. There is evidence for this in the finding of significant brain-weight effects of 21 days of Hebb-Williams maze training in previously isolated, but not in previously enriched, subjects (Cummins et al., 1973). Studies of amphetamine toxicity are also consistent with this hypothesis. Animals which are placed in groups for short periods (hours) are more sensitive to amphetamine toxicity than are short-term isolates as might be expected if the sensory and amphetamine arousal are additive. However, in long-term (weeks) differentially reared subjects, it is the isolates who are more sensitive, as is consistent with their stimulus deprivation induced neurostatic hypersensitivity and hyperarousability (Chance, 1946; Moore et al., 1965; Fuller and Hines, 1967).

## Other Arousing Factors

If the hypothesis of arousal reaction mediating brain modification is correct, then additive and similar brain changes might be expected from other factors which modify arousal. In accordance with this are the drug studies carried out in the Berkeley laboratories (Bennett et al., 1973; Rosenzweig et al., 1971), in which amphetamines or barbiturates were given prior to two hours of daily exposure to the enriched environment. The effects of amphetamines are to increase the magnitude of the arousal reaction and to delay habituation; the effects of barbiturates are the opposite. It might therefore be anticipated that these drugs would enhance and depress, respectively, the effects of enrichment, and such was the case. Even methamphetamine given alone was sufficient to elicit small, though significant, brain-weight changes. Supporting behavioral evidence also comes from the finding that isolation enhances amphetamine induced stereotypy (Sahakian et al., 1975) and reduced barbiturate induced sleeping time (Einon et al., 1976), the latter effect also being greater in the dark phase of the diurnal cycle. These findings must be taken into consideration in future studies of the central nervous system effects of drugs which modify arousal, since it may be that the failure to detect neurochemical sequalae of amphetamine administration (Dewar and Winterburn, 1973; Lima et al., 1966) reflects a failure to consider possible brain-weight changes and the use of environments which are not particularly effective in eliciting amphetamine effects. The reduction in brain responsivity to environmental stimulation induced by barbiturates might well play a role in the brain growth retardation observed in young rats treated with phenobarb (Schain and Watanabe, 1976), and suggests that the use of such drugs in humans should be minimized.

Another factor which modifies arousal is the diurnal rhythm. Rats are more active and aroused during the nocturnal phase, and a short complexity exposure at this time is more effective in eliciting brain changes than one during the day. Indeed, the effects of environmental complexity, methamphetamine, and the dark phase would appear to be additive or even synergistic, since it was this combination which yielded the greatest brain-weight changes (more than any factor alone) (Rosenzweig and Bennett, 1972; Rosenzweig et al., 1971). It is also interesting to note that these changes occurred only with grouped subjects and not with isolates given toys, except when given in conjunction with arousal-producing factors such as amphetamine or the dark phase. This enhancing effect of the group situation may well be due to the contagious nature of arousal. It was repeatedly observed that when toys were added to the cage, only one or two animals would be active initially, but that by their influence the whole group would eventually commence very active exploration and give the appearance of marked arousal. Rosenzweig and Bennett (1972) have found that grouping prolongs the activity of subjects given toys. Thus, arousal contagion may be the mechanism by which social rearing synergizes with the provision of toys to induce the observed brain effects.

This finding of a synergistic effect of toys and social stimulation is analogous to recent findings of sensory effects on human development. Handling, verbal stimulation, and face-to-face visual interaction with the human neonate all tend to result in heightened arousal, and have been found to enhance subsequent cognitive development (Ainsworth, 1973; Casler, 1965; Jones and Moss, 1971; Korner and Thoman, 1972; Lewis et al., 1969; Solkoff et al., 1969). Similarly, the variety and kind of objects available to the infant also enhance cognitive development (Yarrow et al., 1972). However, there is evidence to suggest that the provision of toys alone is insufficient, and that it is only when they become incorporated into a social interaction that they exert significant behavioral effects (Clarke-Stewart, 1974; Williams and Scarr, 1971). Also at the human level, reafference and stimulus contingency would appear to be important in determining the efficacy of a stimulus on cognitive development. Thus, the more contingent the mother's response to her infant's behavior, the higher the IQ score and the faster the development of both object and person permanence (Nelson, 1973; Yarrow et al., 1972).

That a specific modality may subserve environmental effects was originally thought possible, with vision appearing the most likely candidate in view of the marked changes in visual cortex. However, the induction of enrichment-deprivation effects in blind, dark reared, and anosmic subjects (Bennett et al., 1964, 1973, 1974; Rosenzweig et al., 1969) indicates that this is probably not so. However, this finding is consistent with an arousal theory, since the stimulus characteristics which elicit an orientation reaction are largely independent of the sensory modality (Lynn, 1966).

Another factor mediating arousal is reafferent stimulation, and it is interesting to note that visual patterning and toys outside the cage do not appear to be particularly effective in eliciting complexity-like effects in brain anatomy and chemistry (Maki, 1971; Rosenzweig and Bennett, 1977). This suggests that active exploration and interaction with the toys is important in eliciting effects, and this points to the importance of visual and motor feedback (reafference) as opposed to simple afferent stimulation. This concept derives considerable support form three research areas: the role of muscle activity in maintaining arousal (Kleitman, 1963), Prescott's (1970a) emphasis on the importance of kinesthetic input during development, and the many studies (reviewed by Ganz, 1975) demonstrating the important role of reafference in avoiding sensory deprivation effects.

## A PROPOSED MECHANISM FOR AROUSAL TRANSDUCTION

As mentioned, one of the properties of the arousal reaction is a marked increase in electrocortical activity. It now appears that such an increase in neuronal activity could set in motion a heightened metabolism of not only the neurons themselves, but also of glia. It is this heightened metabolism which might be expected to result in an increased net synthesis of cellular consituents, and the consequent development of brain changes.

The following is an hypothesized sequence of cellular events based on a genetic derepression model, by which this bioelectrical-synthetic transduction might occur as an automatic consequence of neuronal activity. As such, it might well be expanded to cellular or molecular models of learning, though the latter might place greater emphasis on additional synaptic events (Kosower, 1972).

The initial neuronal step would seem to be synaptically mediated. Excitatory postsynaptic potentials are followed by alterations in incorporation of RNA precursors in the neurons themselves (Gisiger, 1971; Kernell and Peterson, 1970; Peterson and Kernell, 1970) and adjacent glia (Geinismann, 1971). This neuronal sensitivity to postsynaptic potentials presumably affords the neuron a greater metabolic sensitivity to presynaptic activity than would be possible if the neurons responded only to a complete discharge. Furthermore, this signal for RNA metabolism appears to come directly from the postsynaptic membrane (Gisiger, 1971). It may involve a nonmediatory function of the neurotransmitter on postsynaptic neuronal metabolism (Doemin, 1968), possibly via activation of adenyl cyclase and protein kinases resulting in conformational changes in synaptic proteins (Glassman et al., 1973), and release of substances from the postsynaptic region. Whatever the precise messenger, the first ribonucleic involvement may be RNA polymerase activation, which may also be the first substance synthesized.

Presumably this RNA polymerase synthesis indicates a genome activation involving a derepression due, perhaps, to an interaction between the postsynaptic released substance and repressor. The RNA concentration would next be expected to rise. This might involve all classes of RNA; or the synthesis might be more specific, involving particularly, perhaps, messenger RNA, in view of Appel et al.'s (1967) evidence of the selective synthesis of this substance in the cortex following initial light exposure. If RNA concentration, and particularly that of messenger RNA, is the rate-limiting factor in neuronal protein synthesis (Appel et al., 1967), then this increased RNA synthesis might be expected to precede an increase in protein synthesis. This time course of rises in RNA polymerase, then RNA, and finally protein has been observed following imprinting (Haywood et al., 1970). The time course of the cerebral weight and RNA concentration changes following exposure to the environmental complexity (Ferchmin et al., 1970; Rosenzweig and Bennett, 1978) fits well the last part of this sequence. Mathies et al. (1973) have suggested that pyrimidine precursor availability may be the rate-limiting factor for RNA synthesis. If RNA concentration is in turn a rate-limiting factor for neuronal metabolism, then perhaps precursor administration would augment environment stimulation effects (Walsh and Cummins, 1976a). The formation of mitochondria has been found to be a common response to heightened energy demands and so may well occur in stimulated brains.

Glial changes presumably reflect glial-neuronal metabolic coupling. Certainly potassium efflux occurs from discharging neurons, and glial membrane potential and metabolism are highly sensitive to extracellular potassium levels (Orkand et al., 1966), so that it seems possible that this mechanism could elicit the environmentally induced glial changes. Similarly, ionic fluxes and glutamate release may underlie the initial swelling of dendritic spines after tetanic stimulation (Van Harreveld and Fifkova, 1975).

This postulated sequence of cellular events may be considered a general model for neural response to stimulation. With it, a variety of the observed changes, such as perikaryonal size and RNA concentration, can be explained. However, the cellular mechanism involved in the remainder of this proposed sequence, such as those regulating the more precise and discrete changes such as synaptic morphology, remain beyond the realm of our present knowledge.

## TESTS OF THE AROUSAL HYPOTHESIS

Arousal might be varied by a number of factors and their neural consequences estimated. Factors so used could include the following:

1. The modification of arousal via the habituation process. If the transient nature of the complexity-deprivation difference is to be attributed at least in part to habituation, then any factor which reduces habituation may also be expected to increase the magnitude and duration of the cerebral response to complexity. Such factors may include stimuli to which habituation occurs slowly, or the use of drugs which reduce habituation (e.g. parachlorphenylalanine; Connor et al., 1970).
2. The selection of subjects based on their endogenous arousability prior to the differential rearing conditions. One would predict a positive correlation between arousability and the magnitude of the environmentally induced change. Selection for arousability could be based on either behavioral or physiological parameters.
3. The adoption of certain surgical procedures for arousal modification. For example, lesions in the raphe nucleus have been reported to result in behavioral and EEG evidence of heightened arousal (Kostowski et al., 1968). Once again, those subjects with greater arousability may be more susceptible to environmentally induced cerebral changes.
4. The use of extreme conditions to exceed the peak of the inverted U. If the inverted U relationship, which is known to exist between arousal level and behavioral performance, is generalizable to the cellular level, then it might be anticipated that extremes of stimulation and arousal would be associated with less effect than environmental complexity. It is not known at the present time whether an environment of sufficient complexity could be designed to exceed central nervous system metabolic capacity, and the peak of the inverted U. However, with the combination of arousal-inducing agents, this might be possible. For example, when using amphetamines in differentially reared subjects, one would expect that with increasing dosages the cerebral changes in complexity-reared subject would peak and then decline. Indeed, it might be that the failure to find typical effects in amphetamine-treated older rats (Rosenzweig, 1978) is partially due to exceeding the peak.

## IMPLICATIONS OF THE AROUSAL-INDUCTION HYPOTHESIS FOR THIS AND OTHER FIELDS

It would seem possible that the concept of arousal mediation might be extended to fields other than brain changes, following environmental complexity, social-rearing, and isolation. Whenever the variety and novelty of stimuli are used to induce brain changes, arousal mediation may occur. Such occasions where brain changes analogous to those of complexity-deprivation have been found include normal lighting versus dark rearing

(Bennett et al., 1964; Cragg, 1977; Fifkova, 1968; Fifkova and Hassler, 1969; Gyllensten et al., 1965; Rose, 1967), electric shock (Gardner et al., 1970), multimodality stimulation (Schapiro and Vukovich, 1970), and early handling.

The so-called early handling studies have demonstrated a wide variety of behavioral and endocrine effects following infantile stimulation, though as yet only a few anatomical or biochemical brain parameters have been examined (Altman et al., 1968a, 1968b; Tapp and Markowitz, 1963). Reviews include Schaeffer (1968), Denenberg (1969; Denenberg et al., 1968), Levine (1969a, 1969b), Thompson and Grusec (1970), Russell (1971), and Daly (1973). However, despite over 20 years of vigorous research, the question of how early handling effects are mediated is still a matter of conjecture (Daly, 1973; Russell, 1971; Schaeffer, 1968).

Russell (1971) has reviewed the evidence relating to possible "mechanisms," and subdivided the hypotheses into four categories: (a) direct action of the sensory stimuli on the infant, (b) hypothermia, (c) alteration in maternal behavior, and (d) stress.

He concluded that there is not as yet sufficient evidence to decide between these alternatives, and remained doubtful that a conclusion is possible until physiological mechanisms are identified. In light of this, it is interesting that of the above "mechanisms," the first three actually represent attempts to identify the precise stimuli which elicit the observed effects. Of these, hypothermia has received the most support as a crucial variable, although it is certainly not the only one (Haggett and Werboff, 1968; Hess et al., 1969; Thoman and Levine, 1969) and even it may involve maternal mediation (Barnett and Walker, 1974).

Schaeffer (1968) has used the evidence of the involvement of hypothermia to suggest a basic error in conceptualization of early handling effects. It was long assumed that these changes were mediated by the effects of direct sensory stimulation on the brain, whereas Schaeffer has now hypothesized alterations in metabolism which are independent of central nervous system stimulation. While this is possible, it may also be true that somatic temperature change itself affords central nervous system stimulation.

Certain findings are in accordance with this suggestion, including those of Hutchings (1967), who concluded that the rate of heat loss was more important than the duration, and that longer maintenance of constant level of hypothermia was without additional effects. The importance of rate, as opposed to total amount, of heat loss is consistent with a sensory stimulus mediation model, since the CNS functions as a detector of stimulus change whose sensitivity varies with the rate of change.

Furthermore, there is evidence suggesting that the relationship between amount of cooling and subsequent emotionality is of an inverted U type (Hutchings, 1967; McIver et al., 1968). Several previously puzzling findings would then be explicable using the arousal hypothesis, since almost any form of experimental manipulation would present a host of major novel stimuli against the background of the relatively stimulus-poor laboratory nesting cage. Thus, the mere act of removing an animal from the litter would be expected to elicit nearly maximal arousal, so that further stimulation (such as electric shock or shaking) might cause comparatively little further arousal or, if sufficiently strong, might even reduce it (Zagorylko and Sollertinskaya, 1958). In either case, the expected result would involve very little change above that induced by simple removal from the litter, and this has in fact been found (Levine et al., 1956).

Despite the long history emphasizing the role of stress in mediating handling effects, the evidence seems far from clear. Once again, the central assumption has been that infantile stimulation studies are stressful, an assumption based largely on observation of adrenal activity. Therefore, the handling effects have been ascribed to infantile stress. However, it is becoming increasingly evident that adrenal activity reflects the "mean level of environmental stimulation" (Welch, 1965), and not just noxious stimulation. Therefore, it does not necessarily follow that all the infantile stimulation techniques employed are stressful (aversive, negatively reinforcing), nor that the physiological mechanisms which mediate stress mediate the effects of infantile stimulation.

Levine (1969a, 1969b) has suggested what has become a fairly widely accepted hypothesis; namely, that infantile stress serves to immunize the animal against stress in later life, via the resetting of a pituitary-adrenal hormonostat. However, with the role of stress very much in doubt and the possibility that nonstressful stimuli may also be effective, it seems that the stress immunization hypothesis may well be incorporated within a more general hypothesis of arousal homeostasis. Within this theoretical framework, one would predict that an organism stimulated, and therefore aroused, in infancy by either aversive or nonaversive stimuli would be less aroused, and therefore less stressed, by stimulation in later life. Future studies might test the arousal hypothesis for early handling effects, using the same principles outlined in the section on environmental complexity, and might also attempt to determine the aversiveness of various stimuli to neonates in order to differentiate between the stress and arousal hypotheses.

## CONCLUSIONS

The reader should please note that despite the emphasis on arousal in this chapter, it is not intended to suggest that this is the only mechanism involved in environmental effects. Other mechanisms, such as learning, are almost certainly involved, since an altered arousal state must have as its consequence an altered level of sensory awareness and probability of engram storage. The final picture, therefore, will probably be one of multiple interactive mediating mechanisms, perhaps with arousal as a central process.

One advantage of an arousal hypothesis is that it may afford a degree of integration between concepts of neural and behavioral adaptation to a changing environment. An arousal hypothesis has been frequently advanced to account for certain of the behavioral concomitants of both sensory underload and overload (Konrad and Melzack, 1975; Welch, 1965; Zuckerman, 1969). The present paper extends this hypothesis from the behavioral to the neural level, and allows one to predict certain neural-behavioral correlates, such as the behavioral hyperactivity and heightened susceptibility to brain change apparent in an emergent isolate (Cummins et al., 1973).

One might also predict that sensory overload sufficient to produce cognitive deterioration in animals, normal humans (Gottschalk et al., 1973; Ludwig, 1972), or schizophrenics (Wing et al., 1972) might be accompanied by central anatomical and chemical evidence of metabolism having exceeded the peak of the inverted U curve. Furthermore, since certain schizophrenics seek to maintain sensory input levels within a narrow range (Chapman, 1966; Ludwig and Stark, 1973), it might be expected that this peak would be exceeded more easily than in normals.

The aim of this chapter has been to present arousal mediation as a central process, providing, *by its interaction with other processes*, a common factor necessary for the induction of environmentally induced brain changes. For the sake of simplicity the nature of mechanisms has been discussed in traditional terms. The next chapter, however, contains a deeper analysis of the nature of causality and mechanisms and calls many of our usual assumptions into question.

# CHAPTER 9

# The Tao of Brain: Suggestions of Cross Disciplinary Parallels

> Environment is a function of the consciousness that perceives it.
>
> (Ram Dass, 1975 p.9)

> The history of science is rich in the example of the usefulness of bringing two sets of techniques, two sets of ideas, developed in separate contexts for the pursuit of new truth, into touch with each other.
>
> (Oppenheimer, 1954 p.96)

## Basic Hypothesis

The basic hypothesis of this chapter is that as perceptual sensitivity increases beyond a certain threshold we penetrate beyond the realm of our ordinary perception of the world and its concomitant "reality" and obtain a fundamentally different view of nature. This heightened perception may be obtained through direct perceptual training such as in meditation, or indirectly by augmentation and systematization through instrumentation and experimentation as in advanced science. But, however it is obtained, perceptual enhancement of sufficient degree may reveal a different order of reality from that to which we are accustomed. Furthermore, the properties so revealed will be essentially more fundamental and veridical than the usual, and will display a greater degree of commonality across disciplines. Thus, as empirical disciplines evolve and become more sensitive they will uncover phenomena and properties which point toward underlying commonalities and parallels between disciplines. It appears that the neurosciences may now be at the threshold of this evolutionary stage. In order to evaluate this hypothesis of emerging cross disciplinary parallels, it is necessary to range somewhat far afield from basic brain research. The nature of scientific paradigms, the frontiers of modern physics, and the perennial claims and recent research in the consciousness disciplines must all be considered. It is hoped that the reader will bear with this broad range of areas and the boldness of the hypothesis.

## Paradigms and Paradigm Clash

This concept essentially represents a new paradigm. A paradigm is a kind of "super theory," a theory or formulation about the nature of reality of such scope that it is capable of encompassing a major portion of the known phenomena in its field. Copernican or heliocentric astronomy, the idea that the planets revolve around the sun, is an example of a paradigm that still guides astronomy. Now, supposedly all scientific theories, including paradigms, are open-ended and continually accessible to modification or refutation. But in point of fact a variety of psychological factors and inertia tend to reduce this flexibility. Because of their very success they eventually tend to be taken for granted and to insinuate themselves into the researcher's psyche in such a way that these "normative paradigms" (Wilson, 1970) become implicit, unquestioned conceptual frameworks and filters which supply the "natural" and "sensible" way of looking at things. As Charles Tart (1975b) has commented, "once a paradigm becomes implicit, it then acquires a tremendous controlling power over its adherents," and it may even seem that the paradigm is obviously correct because "this is the only way it could be" (Rychlak, 1968; Scriven, 1969). In fact, we may only recognize that we are being guided by a paradigm when a new one emerges. At this stage a new paradigm becomes difficult. New paradigms tend to be broader and even metaparadigmatic. It is typical of new paradigms that they initially seem illogical if not downright ridiculous, simply because they run counter to our previous (paradigmatic) way of thinking about things. Thus, when a new paradigm begins to emerge we have what Thomas Kuhn (1970) calls a "paradigm clash."

What this chapter hypothesizes is that we may be witnessing such a clash in which one of the most fundamental paradigms of Western science, the classical Greek concept of the universe as essentially atomistic, divisible, isolable, static and nonrelativistic, is in the process of replacement, not just for physics but for all science. Scientists first suggested the new paradigm of the universe as holistic, indivisible, interconnected, and relativistic, in response to the unexpected findings of modern physics. What is being suggested here is that the new paradigm is applicable to the psychological and neurosciences, and ultimately to all of science, and that as the individual branches of science evolve we may find evidence of cross disciplinary convergence, not only between sciences but also between science and the consciousness disciplines as each discovers a similar fundamental, underlying substrate of reality.

## Perceptual Limitations

Both modern science and ancient consciousness disciplines point out that our usual perception is limited and distorted to a degree that we do not recognize. For millenia the meditative and yogic disciplines have devoted themselves specifically to this problem and have stated that it is only when we begin working to augment our perceptual sensitivity and accuracy that we begin to appreciate the existence and magnitude of the problem. Although it is often forgotten, the aim of modern science is essentially the same: namely, to transcend our usual perceptual limitations in order to obtain a more accurate and sensitive picture of the universe. Western psychology has long recognized and explored perceptual limitations, but in recent years certain data derived from physics has begun to confirm both the limitations of perception and the picture of underlying reality described by the meditative-yogic disciplines. They suggest that our usual perceptual limitations tend to produce consistent but unrecognized distortions. These distortions include tendencies to solidify, dichotomize, separate, oversimplify, and concretize, and to underappreciate the extent of continuous flux, impermanence, interconnectedness, and holistic consistency of the universe. Thus our implicit models of the universe have been essentially those of certain of the Greek philosophers who, on the whole, viewed nature as essentially static, atomistic and nonrelativistic. Thus both the consciousness diciplines and modern physics suggest that these distortions are so pervasive and unrecognized that our usual picture of the universe (i.e., "reality") is fundamentally erroneous or illusory. The word "illusory" has often been misunderstood to imply that the world does not really exist: Rather, it simply means that our perception of it is colored and distorted to an unrecognized degree.

When perceptual limitations are overcome, the newly uncovered reality appears strikingly different from the everyday one. In general, the following characteristics are descriptive of the reality thus perceived: The universe appears to be *constituted of a single common substrate* ("of single essence"; Sengtsan, 1975), *nondualistic* as opposed to dichotomous, a *unitive whole* as opposed to unrelated parts, *interconnected* as opposed to comprised of separate and isolated components, *in continous motion* or flux as opposed to static, *impermanent and ephemeral* as opposed to lasting and permanent, *empty* (largely constituted by non-solid empty space) rather than solid, *infinitely complex* as opposed to comprehensible in terms of a finite number of dimensions, *acausal* (but not anticausal), i.e., transcending cause and effect, rather than comprehensible within traditional models of causality, *foundationless and self-consistent* (bootstrap theory) in that no component or mechanism is ultimately more fundamental than another, and hence inexplicable in terms of a limited number of fundamental mechanisms, *statistical and probabilistic* rather than certain, *paradoxical* rather than fully intellectually comprehensible, codifiable, and communicable, *inextricably linked with consciousness* rather than separable into consciousness and objects of consciousness: "The world may be called physical or mental or both or neither as we please; in fact the words serve no purpose" (Bertrand Russell, cited by Wilber, 1977, p.38).

Thus the fundamental ontology which is being revealed is largely empty, dynamic, fluid, impermanent, holistic, interconnected, interdependent, foundationless, self-consistent, paradoxical, probabilistic, and infinitely over-determined, and is a function of consciousness. This description and the sources from which it is derived will be discussed in more detail subsequently.

## Convergence of Disciplines

We might hypothesize that since the above description refers to fundamental properties common to all phenomena, perception of sufficient sensitivity and veridicality will begin to recognize these properties no matter what the perceptual mode and object of perception. Thus, any object, if examined by any perceptual mode with a sensitivity sufficiently enhanced, either by direct training or by scientific instrumentation, might present a picture of its inherent nature as described above.

But at this level of greater sensitivity another factor enters in, namely, the consciousness of the observer. As will be shown subsequently, the picture of the universe that emerges is a function or creation of consciousness; observer and observed ultimately cannot be separated. Thus, any discipline will begin to detect fundamental properties of both the objects under investigation and the observational system, including the consciousness of the observer, and these will display the characteristics discussed above, and thus transcend disciplinary boundaries and point to cross disciplinary commonalities. Beneath the initial appearance of infinite diversity may be found a complementary, underlying, essential commonality pervading all of nature, which was first described centuries ago, and which transcends traditional disciplinary boundaries.

To date the most exquisite refinement of perception has apparently occurred in meditation and in physics. Thus, in view of the above hypothesis, it is not surprising that modern physics has been involved in a revolution which has shaken the very foundation of science. For the reality revealed, especially that of the subatomic level, is so discordant with our usual picture of reality, so paradoxical, so downright bizarre, as to defy description in traditional terms and to call into question some of the most fundamental assumptions of Western science and philosophy. Yet these same findings, which do not fit at all with our usual descriptions of reality, are strikingly reminiscent of those descriptions given repeatedly across cultures and centuries by advanced practitioners of the consciousness disciplines. Indeed, much of modern physics can be viewed as a rediscovery of ancient wisdom (Oppenheimer, 1954; Bohr, 1958; LeShan, 1969; Capra, 1975, 1976; Wilber, 1977; Beynam, 1978).

> The general notions about human understanding . . .which are illustrated by discoveries in atomic physics are not in the nature of being wholly unfamiliar, wholly unheard of, or new. Even in our own culture they have a history, and in Buddhist and Hindu thought a more considerable and central place. What we shall find is an exemplification, and encouragement, and a refinement of old wisdom
>
> (Oppenheimer, 1954 p.89)
>
> For a parallel to the lesson of atomic theory . . . (we must turn) to those kinds of epistemological problems with which already thinkers like the Buddha and Lao Tzu have been confronted, when trying to harmonize our position as spectators and actors in the great drama of existence
>
> (Bohr, 1958 p.20)

Fritjof Capra (1975) suggests in his book The Tao of Physics, which describes the parallels between modern physics and ancient mysticism (a much misunderstood and misused term which actually refers to disciplines embodying specific types of perceptual training):

> why we might expect the world views of modern physics and Eastern mysticism to be similar. Both emerge when man inquires into the essential nature of things-into the deeper realms of matter in physics; into the deeper realms of consciousness in mysticism-when he discovers a different reality behind the superficial mechanistic appearance of everyday life.
>
> The parallels between the views of physicists and mystics become even more plausible when we recall the other similarities which exist in spite of the different approaches. To begin with, their method is thoroughly empirical. Physicists derive their knowledge from experiments; mystics from meditative insights. Both are observations, and in both fields these observations are acknowledged as the only source of knowledge.

In general, we might expect that the more sophisticated and hence sensitive a discipline or branch of science, the more it will begin to unearth and point towards underlying phenomena and properties which are in fact universal. This will not be instead of, but in addition to, the unique properties of the specific objects which it studies. Perhaps the same principles and truths will be rediscoverable at many levels, with many degrees of subtlety and pervasiveness. The more sensitive the science, the more these underlying principles will be recognized as inherent to the nature of the objects of study of that science. We might use the analogy of the ocean and its waves, since this is an analogy which the meditative disciplines have often employed. The waves on the surface have many forms, each has its own unique configuration and patterning; but as we go deeper into any one of them, we begin to discover its common features with others, and deeper still we find ourselves exploring the ocean itself no matter which individual wave we began with. We now become aware of the fundamental, underlying commonality of all waves, and that this commonality may manifest itself in an infinite variety of forms. Although we might start from different points and take any one of an infinite variety of waves as our object of examination, once we had passed deeply enough, we would find ourselves describing the same fundamental reality without in any way negating the uniqueness and validity of each apparently separate and distinct wave. Our experience of the nature of waves would change as a function of the shift in our perception. Thus the degree to which a discipline decribes this underlying commonality may provide an index of its sensitivity and maturity.

Clearly at the present time physics, one of the oldest of the natural sciences, has progressed furthest and deepest. The biological sciences. which to a large degree are dependent for their advances on technologies provided by the physical sciences, have not come as far towards a recognition of an underlying reality and commonality. However, it is the hypothesis of this chapter that the neural and psychological sciences may have reached a point now at which a shift in paradigms similar to that occurring in physics needs to be considered, to provide holistic, dynamic and probabilistic models of brain and brain function. For the neurosciences have as their object of study the most complex and sophisticated receptor known to exist. In examining its responses to stimuli we can recognize that these indicate underlying interdependence with the environment and similar fundamental properties to those described above. Thus the psychological and neuro-sciences, and especially the ecology of brain, can begin to glimpse the underlying ocean. In doing so, they may begin to appreciate the underlying commonality and convergence with ancient mysticism and modern physics and the potential convergence of all the empirical disciplines. The following sections will examine the data for such convergence, after giving a brief overview of the consciousness disciplines.

## The Consciousness Disciplines

It is little known that behind the great Eastern religions such as Buddhism and Hinduism lie highly sophisticated psychologies and cartographies of consciousness (Tart, 1975b; Wilber, 1977; Goleman, 1977, 1978; Walsh and Shapiro, 1978a, 1978b; Walsh and Vaughan, 1978). These psychologies, which are variously called consciousness, meditative, yogic, mystical, and spiritual traditions, all assert that our usual perception and state of consciousness are suboptimal and illusory and that through intense mental training such as in meditation or yoga it is possible to transcend these limitations and obtain profound insight into the nature of consciousness, reality and self. The final transcendence of these limitations is called by such names as enlightenment, liberation, and moksha. It is stated that with the clear perception which accompanies this condition, the universe appears as described above, and that one can look back now and recognize the perceptual distortions which existed prior to enlightenment.

The depth and significance of these disciplines has clearly been underestimated by Western science until very recently. This probably reflects cultural factors, the fact that some experiential familiarity with the phenomena they describe seems essential for a full intellectual understanding, and the importance of state dependent learning. State dependent learning is a recently recognized phenomenon in which material learned under one state of consciousness may not be retained or understood under another (Tart, 1975a). The consciousness disciplines are closely involved with altered states, and hence understanding a significant proportion of their material involves state dependency and may necessitate some direct experience of the disciplines. Once this is recognized, what formerly seemed arcane and even nonsensical may become comprehensible and even highly sagacious. These disciplines are now beginning to come under intense investigation and to reveal a depth of wisdom which was unsuspected only a few years ago (for reviews see Tart, 1975a, 1975b; Wilber, 1977; Goleman, 1977, 1978; Shapiro and Walsh, 1978; Walsh and Shapiro, 1978b; Walsh and Vaughan, 1979, 1978; Thetford et al., 1978).

## THE ECOLOGY OF BRAIN AND THE ALTERNATE PARADIGM

The following sections will attempt an examination of the ecology of brain structure and function in light of the alternate paradigm suggested by physics and the consciousness disciplines. This will be done by examining brain research data relevant to certain dimensions which have been found to characterize the alternate paradigm, e.g., holism, context, effects of the observer, dynamism and impermanence, probabilism and uncertainty, causality, and limits of conceptualization and communication.

It should be noted that it is not being suggested that exactly the same laws necessarily apply across different realms, for example, from quantum physics to the macrosopic world. Whether they do or not is an extremely complex and unresolved question taxing some of the best minds of both physics and philosophy (Putnam, 1964). Rather it is being suggested that our usual perception provides a picture of reality which is transcended as perception becomes more sensitive. This emergent picture of reality, which is a function of both the universe and of perception itself, will display heretofore unrevealed commonalities.

### Holism

The recognition of the holistic inseparability of phenomena represents one of the fundamental shifts in science of recent years. Thus Bertalanffy (1969: p.45), speaking from one of the more general perspectives of modern science, general systems theory, states:

> We may state as a characteristic of modern science that (the) scheme of isolable units acting in one way causality has proved to be insufficient. Hence the appearance, in all fields of science, of notions like wholeness, holistic, organistic, Gestalt, etc., which all signify that in the last resort, we must think in terms of systems of elements in mutual interaction

Or again, in the words of Tielhard de Chardin (1965: p.43-44):

> Up to now we have been looking at matter as such, that is to say according to its qualities and in any given volume - as though it were permissible for us to break off a fragment and study this sample apart from the rest. It is time to point out that this procedure is merely an intellectual dodge. Considered in its physical concrete reality . . . the universe cannot divide itself, as a kind of gigantic 'atom', it forms in its totality . . . the only real indivisible . . . The further and more deeply we penetrate into the matter, by means of increasingly powerful methods, the more we are confounded by interdependence of its parts. Each element of the cosmos is positively woven from all others . . . It is impossible to cut into this network to isolate a portion without it becoming frayed and unravelled at all its edges. All around us, as far as the eye can see, the universe holds together, and only one way of considering it is really possible, that is, to take it as a whole, in one piece.

This essential unity and inseparability of parts has been one of the fundamental and recurring discoveries of particle physics, a discovery which is a rediscovery of principles enunciated millenia ago by the explorers of consciousness. Thus Bohm and Hilley (1975; p.96-102) note that:

> One is led to a new notion of unbroken wholeness which denies the classical idea of analyzability of the world into separately and independently existing parts . . . we have reserved the usual classical notion that the independent "elementary parts" of the world are the fundamental reality, and that the various systems are merely particular contingent forms and arrangements of these parts. Rather, we say that inseparable quantum interconnectedness of the whole universe is the fundamental reality, and that relatively independently behaving parts are merely particular and contingent forms within this whole.

Or in the words of Fritjof Capra (1975: p.118):

> As we study the various models of sub-atomic physics we shall see that they express again and again, in different ways, the same insight that the constituents of matter and the basic phenomena involving them are all interconnected, interrelated and interdependent; that they cannot be understood as isolated entities but only as integrated parts of the whole.

Indeed, many of the statements by particle physicists could be mistakenly identified as coming from the writings of the great mystics. For example:

> An elementary particle is not an independently existing unanalyzable entity. It is, in essence a set of relationships that reach outwards to other things.
>
> (Stapp, 1971: p.1310)

> The world thus appears as a complicated tissue of events, in which connections of different kinds alternate or overlap or combine and thereby determine the texture of the whole.
>
> (Heisenberg, 1958: p.107)

In essence, then, quantum physics forces us to discard the classical concepts of independent, isolable units whose structure and function can be investigated and determined independent of the remainder of the universe. Rather we must replace these concepts with a vision of an interconnected and interdependent web of relationships among the parts of a unified and indivisible whole. And this image is not restricted to, or derived solely from, the subatomic world, but is again forced upon us when we change scale and examine the macroscopic, for in cosmology the same patterns emerge. In the words of astronomer Fred Hoyle (1960: p.304):

> Present day developments in cosmology are coming to suggest rather insistently that everyday conditions could not persist but for the distant parts of the universe, that all our ideas of space and geometry would become entirely invalid if the distant parts of the universe were taken away. Our everyday experience even down to the smallest details seems to be so closely integrated to the grand-scale features of the universe, that it might be well-nigh impossible to contemplate the two being separated.

Thus, for example, according to the physicist and philospher Ernst Mach, the inertia of an object is not an intrinsic property but is rather a measure of its interaction with the rest of the universe. In Mach's view, matter has inertia only because there is other matter in the universe. Indeed this concept of inertia, which has become known as the Mach Principle, provided part of Albert Einstein's original motivation for constructing the general theory of relativity (Capra, 1975). It is only in the everyday "middle range" scale of things that we are able to separate parts and consider them in isolation, and even then this constitutes only a first approximation. Whenever we change focus, either to the very large or the very small, the inherent unity of the universe forces itself into our awareness.

This fundamental unity is one of the basic tenets of the consciousness disciplines. Concepts, they hold, are inherently dichotomous. One cannot have a concept "up" without a diametrically opposite concept "down", nor "big" without "little". When we see the world through the filter of concepts we necessarily see a dichotomous universe. It cannot be otherwise since thought and intellect cannot be otherwise, and thus ordinarily we misperceive the true nature of things without recognizing misperception. Thus the journey of awakening, the task of enlightenment, the spiritual work, the training of perception, the expansion of awareness, the coming into the Tao, liberation, freedom, God realization, transcendence, whatever it is called and wherever it is practiced, aims at freeing awareness from the usually unrecognized distortions of our conditioned conceptual filters. The end result is held to be the recognition of the true unitary nature of one's self and the universe. Thus in the words of the third Zen patriarch (Sengtsan, 1975), written millenia ago:

> Do not remain in the dualistic state:
> avoid such pursuits carefully
> If there is even a trace
> Of this and that, of right and wrong,

> The mind-essence will be lost in confusion. . .
> When thought objects vanish, the thinking-subject vanishes,
> As when the mind vanishes, objects vanish.
> Things are objects because of the subject (Mind):
> The Mind (subject) is such because of things (objects).
> Understand the relativity of these two
> And the basic reality; the unity of emptiness.
> In this emptiness the two are indistinguishable
> and each contains in itself the whole world.
> All dualities come from ignorant inference.
> They are like dreams or like flowers in air;
> Foolish to try to grasp them . . .
> If the Mind makes no discriminations.
> The ten thousand things
> Are as they are, of single essence.
>
> To understand the mystery of this one essence
> is to be released from all entanglements.
> When all things are seen equally
> The timeless Self-essence is reached.
> No comparisons or analogies are possible in this causeless,
> relationless state
> When such dualities cease to exist
> Oneness itself cannot exist.
> To this ultimate finality
> No law or description applies . . .
> In this world of Suchness
> There is neither Self or other-than-self
> To come directly into harmony with this reality
> Just simply say when doubt arises, 'Not two'.
> In this 'not two' nothing is separate
> Nothing is excluded.
> No matter when or where,
> Enlightenment means entering this truth

Even such fundamental relativistic tenets of modern physics as the inherent unity of space and time were foreshadowed centuries ago by the explorers of inner space. From the state of consciousness called by such names as enlightenment, and liberation, this inseparable unity is directly experienced. "Entering into the samadhi of purity (one obtains) all penetrating insight. That enables one to become conscious* of the absolute awareness of the universe" (Ashvaghosha, 1900: p.93). Indeed some of these descriptions might easily be mistaken for modern physics texts:

> We look around and perceive that . . . every object is related to every other object . . . not only spatially, but temporally . . . As a fact of pure experience, there is no space without time, no time without space; they are interpenetrating (Suzuki, 1959: p.33).

Compare this with the description of space-time first introduced by Herman Minkowski in 1908:

---

* Samadhi is a general term denoting a state of consciousness marked by extreme concentration and perceptual clarity.

> The views of space and time which I wish to lay before you have sprung from the soil of experimental physics, and therein lies their strength. They are radical. Henceforth space by itself, and time by itself, are doomed to fade away into mere shadows, and only a kind of union of the two will preserve an independent reality (Einstein, 1923: p.75).

Thus, at their most fundamental levels, both physics and the consciousness diciplines point towards a holistic view of the universe.

Likewise, studies of the ecology of brain suggest a similar holistic, interconnected unitive view. The brain and the remainder of the universe constitute a coherent whole which cannot be divided without constituting an artificial and distorting duality which hides their underlying unity and interconnectedness. The structure and function of the brain are a function of the whole and of the interdependence of the parts of this whole. The record of their interaction is dynamically engraved in chemical and anatomical script in the neural pathways. The universe comes to know itself through the brain, and within its' limits, the brain modifies and adapts itself to the demands of the universe.

It is, of course, impossible to completely examine one without the other. But in our usual explorations we forget this, or, at the very least, choose to ignore it. Still, as our investigations become more sophisticated, there comes a point at which this ignorance can no longer be maintained in the light of increasingly intrusive confounding variables and results. At this point, as so often happens, yesterday's confounding variable becomes today's independent variable. Thus, for example, early studies of animal behavior, brain structure and function, effects of drugs, hormones, genetics, and nutrition did not consider as important factors such as social interaction, group size, lighting, and handling; and instead relegated these to the limbo of insignificant maintenance routines. Now, however, they are recognized as capable of inducing significant and long-lasting changes in both brain and behavior, and it seems likely that any stimulus will cause effects within the brain. To fail to recognize and control for this may reduce experimental sensitivity and validity as well as maintain our inappropriate atomistic models. The effects, though possibly minute and, from the perspective of an observer, lost within neural noise, will still be engraved within the brain, an embodiment of the history of the brain-environment interaction which will, even though minutely, forever exert an influence upon behavioral and functional tendencies. It might be argued that the structure, behavior, and function of the environment penetrate into, and are embodied within, the structure and function of the brain. In the words of Murphy and Spohn (1968):

> The structure of the environment, both its temporal and spatial structure, is built within the person.

The very raison d'etre for the brain is to know, record and react to the remainder of the universe. To consider them as ultimately separate is clearly artificial. As in physics and the consciousness discipines, the evolution of the psychological and neuro-sciences leads to a unitive, holistic perspective.

## Understanding Phenomena in Terms of Context

As a general principle, it seems that the structure, function, and properties of any entity can only be understood in the fullest sense when viewed contextually. Viewed independently, an object has no significance or meaning. Rather, these are provided by the context which holds it or from which it is viewed (Erhard et al., 1978).

Thus, to understand neural structure and function fully we need to examine the brain when it is engaged in meaningful activity, function and behavior, and in terms of its relationship to its environment. The necessity for this examination of activity within an ecological or contextual approach has been demonstrated repeatedly throughout this book. For example, studies of the effects of malnutrition and other forms of brain damage lead to less sensitive, less valid, and less ecologically relevant data unless the nature of the environment and the organism's interaction with it are taken into consideration. Thus, the malnourished child or the brain damaged infant may prove especially sensitive to forms of sensory deprivation (Richardson, 1976; Levine and Wiener, 1976; Greenough et al., 1976). It is important here to remember the interactive and often non-additive nature of the determination of brain structure, function, and behavior (Walsh and Cummins, 1976b). Thus we cannot assume that we can always extrapolate from studies of isolated brain to situations in which the brain is responding to either sensory environment or to other agents. Only through the brain's interaction with its environment and with the observer can we obtain a functional measure of brain activity.

Once again, this recognition is remarkably similar to those of the physicist, the general systems theorist, and the mystic, all of whom have found their subject matter to be understandable only in terms of its interaction and relationship. Thus, Niels Bohr, one of the founders of atomic theory, commented as early as 1934 that "isolated material particles are abstractions, their properties being definable and observable only through their interaction with other systems" (p.57).

Capra (1975: p.69) also gives a nice description of this finding:

> The properties of a particle can be understood only in terms of its activity - of its interaction with the surrounding environment . . . the particle, therefore, cannot be seen as an isolated entity, but has to be understood as an integrated part of the whole.

Similarly for general systems theorist Francisco Varela (1975: p.26): "My ability to see what its properties are is limited by what kinds of interactions I can have with it." Indeed, some interpretations of quantum physics hold that certain properties, such as the position and momentum of a subatomic particle do not exist in any customary form until they are observed (Putnam, 1964).

Similarly, for the explorer of consciousness, humanity is recognized as an integral part of the universe and can be understood only in terms of human-relationship to this universe. To examine humankind independently is not only inappropriate but also impossible, and in the final state of enlightenment the distiction between self and not-self, me and not-me, consciousness and not-consciousness, subject and object, disappear, and there is only the nondual, the universe as a whole, whose inseparable component, humankind, can be understood only in terms of this whole.

## The Observer and the Observed

> What we observe is not Nature itself, but Nature exposed to our method of questioning . . . Natural science does not simply describe and explain Nature; it is part of the interplay between Nature and ourselves (Heisenberg, 1958: pp.81, 107).

Our observations of the universe are not independent, inviolable givens whose nature and values remain fixed regardless of how we determine them. Rather, they represent functions of the observing system and this system includes us, the observers. For reality is a function of the observational systems employed to examine it, and ultimately of human consciousness itself. This is a millennia-old thesis of the consciousness disciplines, a decades-new finding of physics, and a dawning realization of psychology and neuroscience.

As we have previously noted, the trend of research on environmental influences on brain and behavior suggests that any stimulus will elicit effects, although many of these will likely be too small or too subtle for us to detect. This means that any stimulus that occurs in preparation for observation or during the observation process, will affect that which is to be observed. Therefore, our brain measures will reflect not simply the properties of the brain elicited by the independent variable, but also a complex interaction between brain, independent variable, observational procedure, and finally, the observer. As the sensitivity of our experiments increases, especially sensitivity to rapid change, the pervasive existence of such interactions between observer and observed is likely to force itself more and more into our awareness.

Already there is accumulating evidence of this type of interaction in brain research where the importance of the observation procedure per se is being increasingly recognized. For example, it is common to inject substances into animals in order to administer pharmacological agents or trace metabolic activities. However, it is now clear that the very process of injection itself is capable of altering brain weight and presumably other parameters as well, in a manner which interacts with prior experiental history (Ferchmin and Ferchmin, 1973).

This problem may become especially important in the case of measures of rapid biochemical processes. Already we need to be wary of biochemical studies which do not specify the stimulus conditions during the observation-experimentation procedures, and which do not test the effects of these procedures and their interactions with the experimental variables.

This problem becomes especially apparent when we examine behavior. For example, in the case of the open field test which is designed to measure such factors as response to a novel environment, exploration, and emotionality, and is perhaps the most widely used test of animal behavior, a host of variables related to the measurement procedures interact with other variables to determine outcomes (Walsh and Cummins, 1976b). Thus, the procedure by which the animal is removed from its home cage and placed in the open field, the ambient lighting, noise and odors, variations in the position, visibility, and familiarity of the observer all appear to influence the observed behavior. It would be problematic enough if these influences were simple additive ones, but it seems that they are far more likely to interact one with another, and with the genetic and experiential background of the experimental subject. The complexity of the multiway interactions is truly awesome. In addition, the use of different types of apparatus designed to measure the same behavior, such as running wheels to measure ambulation, may yield surprisingly low correlations between apparati.

Interestingly, the effects of the observation process have been specifically examined and employed therapeutically in the field of behavioral self-control. Simple observation of one's own behavior tends to change it spontaneously, often in a therapeutic direction, and this process is known as "reactivity" (Thoresen and Mahoney, 1974). Indeed, this may even represent one of the major mechanisms mediating personality change in people practicing the consciousness disciplines (Shapiro and Walsh, 1978).

In summary, it seems clear that we should not examine brain and behavior without appreciating that what we observe represents a function of the observation process, and ultimately, of ourselves, the observer.

Consider the similarity of this description to the following:

> At the atomic level, "objects" can be understood only in terms of the interaction between the processes of preparation and measurement. The end of this chain of process lies always in the consciousness of the human observer . . . The crucial feature of atomic physics is that the human observer is not only necessary even to observe the properties of an object, but is necessary even to define these properties. In atomic physics, we cannot talk about the properties of an object as such. They are meaningful only in the context of the object's interaction with the observer (Capra, 1975: p.126).

> Nothing is more important about the quantum principle than this, that it destroys the concept of the world as 'sitting out there,' with the observer safely separated from it by a 20-centimeter slab of plate glass. Even to observe so minuscule an object as an electron, he must shatter the glass. He must reach in. He must install his chosen measuring equipment. It is up to him to decide whether he shall measure position or momentum. To install the equipment to measure the one prevents and excludes his installing the equipment to measure the other. Moreover, the measurement changes the state of the electron. The universe will never afterward be the same. To describe what has happened, one has to cross out that old word 'observer' and put in its place the new world 'participator.' In some strange sense, the universe is a participatory universe (Wheeler, 1973: p.244).

Just how fundamental this participatory nature of the reality we perceive is, can perhaps be best demonstrated by an examination of what is called "S-matrix theory." This is a complex mathematical framework which seems to be the most appropriate to date for describing that class of sub-atomic particles called hadrons, and their interactions. Hadrons are especially interesting in view of the fact that they comprise a family of particles no one of which is fundamental, but which rather are comprised of one another and can iteract, combine, and transmute, in a complex but lawful variety of ways (Capra, 1975). The S-matrix is a collection of probabilities for all possible reactions involving hadrons, and in a fully developed theory would be capable of predicting the probabilities of all possible interactions and transmutations, as well as the structure of the individual hadrons; and all this in a dynamic fashion in which each hadron is recognized as an integral component of an inseparable network of particles and interactions. The central aim of S-matrix theory is to derive the structure of the S-matrix in terms of general principles drawn from relativity theory, quantum theory, and causality theory. The implications of such an undertaking are profound. To quote Capra (1975: p.266):

> All three of the general principles are related to our methods of observation and measurement, that is, to the scientific framework. If they are sufficient to determine the structure of hadrons, this would mean that the basic structures of the physical world are determined, ultimately,

> by the way in which we look at the world. Any fundamental change in our observational methods would imply a modification of the general principles which would lead to a different structure of the S-matrix, and would thus imply a different structure of hadrons.
>
> Such a theory of subatomic particles reflects the impossibility of separating the scientific observer from the observed phenomena . . . It implies, ultimately, that the structures and phenomena we observe in nature are nothing but creations of our measuring and categorizing mind.

Such a conclusion is nothing new to the mystic. Indeed, it represents almost a paraphrasing of statements which could be found in texts across mulitiple cultures and centuries (Guenther, 1976; Guenther and Kawamura, 1975; Wilber, 1977). "All is mind," claim the mystics, "we know nothing but our mind, mind and universe are one, everywhere you look you see only yourself, the universe occurs within the mind, mind is boundless, unlimited . . . " Statements such as these, which have always mystified and irritated those of us who have not explored these disciplines, can now begin to make some sense. Consider how exquisitely some of the following quotations might apply to the picture of reality now emerging from modern science.

> The Buddhist does not believe in an independent or separately existing external world, into whose dynamic forces he could insert himself. The external world and his inner world are for him only two sides of the same fabric, in which the threads of all forces and of all events, of all forms of consciousness and of their objects, are woven into an inseparable net of endless, mutually conditioned relations (Govinda, 1969: p.93).
>
> When the oneness of the totality of things is not recognized, then ignorance as well as particularization arises, and all phases of the defiled mind are thus developed . . . All phenomena in the world are nothing but the illusory manifestation of the mind and have no reality of their own (Ashvaghosha, 1900: pp.79, 86).
>
> Out of mind spring innumerable things, conditioned by discrimination . . . these things people accept as an external world . . . what appears to be external does not exist in reality; it is indeed mind that is seen as multiplicity; the body, property, and above - all these, I say, are nothing but mind (Suzuki, 1952: p.242).

Whatever we examine, whatever disciplinary context from which we examine it, whatever observational system we employ, when we look closely enough we recognize that we are describing ourselves and our interaction with the object. Actually, the situation is even more complex, since what we are perceiving is a series of interactions, firstly of the object with the measuring system, then within and among various components of the system, and finally, of the system with the consciousness of the observer.

Reality, than, or the universe as we perceive and describe it, is a function and creation of consciousness, both directly and indirectly via the observational methods, approaches, tools, objects of examination, and analyses we choose to employ. This suggests that there are an infinite number of realities, any one of which will emerge or be created as a function of our state of consciousness, observational system, etc. In the words of Ram Dass (1978: p.21):

> We grow up with one plane of existence we call real. We identify totally with that reality as absolute, and we discount experiences that are inconsistent with it . . . what Einstein demonstrated in physics is equally true of all other aspects of the cosmos: all reality is relative. Each reality is true only within given limits. It is only one possible version of the way things are. There are always multiple versions of reality. To awaken from any single reality is to recognize its relative nature.

Thus it seems that the universe as we know it, and the principles and laws we derive to describe it and ourselves, actually describe our interaction with the universe. We are co-creators of reality, of the laws, structure, and function of the universe. The universe is what it is; what we perceive it to be is a function of us. The laws by which we see it operating, the principles which describe and predict it, are no less than our co-creations. With the collapse of Cartesian dualism and the advent of "bootstrap theory" (which will be discussed in greater detail later), consciousness becomes an inseparable and necessary component of the descriptions of the basic workings of this co-created universe. This is the fundamental message brought to us by those exploring the frontiers of both science and consciousness.

Therefore, if we are to understand the universe we must understand ourselves. To perceive clearly we must first know and take into account our own minds; this is the fundamental goal of the consciousness disciplines. The following are quotations by two eminent explorers of consciousness. The first is by Karlu Rinpoche (1977), a Tibetan monk now living in the United States, and the second is by another Tibetan monk, Tilopa (1977), who lived centuries ago.

> There is a reality behind the senses. You are that reality.

> To see if this be true, look within your own mind.

It seems that in the neurosciences and physics it is becoming increasingly apparent that the consciousness disciplines were essentially correct when that claimed that consciousness and the objects of consciousness could not be separated. What we examine is actually the *process of observation* rather than any detached object of observation, and what we see represents a mutual co-creation of consciousness and its object. Thus we can never speak about nature without speaking about ourselves, both because we are, and because we create, the nature of which we speak. Furthermore, Heisenberg's famous uncertainty principle, which states that, at the subatomic level, the very process of measurement affects what is being measured and limits the information which can be obtained, very well applies across all levels and branches of science.

## Dynamism and Impermanence

> As investigators in brain research we repeatedly refer to the "dynamic" nature of the brain and nervous system, yet we design, and we interpret, our experiments on completely different assumptions. Our assumptions and arguments are mainly Aristotelean. That is, we really don't see brain function as dynamic at all, we see it as a static thing, a thing that can be taken apart one piece at a time. We tend to regard a structure as doing a certain thing, regardless of the conditions we impose upon it, and regardless of its relationship to other parts and to the whole.

> Other sciences, such as physics and chemistry, have found Aristotelian concepts are only partly true and have substituted Galilean logic or the notion of relativity in their scientific thinking. From their point of view an entity does not exist by itself. In determining an entity, the *situation* assumes as much importance as does the object. The object's identity and function may change with the situation (Beck et al., 1969).

Our usual view of the brain is that it is relatively immutable, at least structurally. The changes that do occur are thought to be slow and in response to chronic functional demands. The chemical and electrical picture, however, is very different. Neurons may fire hundreds of times per second, and with this go extraordinarily rapid biochemical shifts and processes which constitute ever new electro-chemical, spatio-temporal fluxes and patternings. Often we cannot accurately describe fixed states, but can talk only in terms of dynamic processes and equilibria which may constitute homeostatic balances. For example, although catecholamine levels may remain stable, this reflects a dynamic equilibrium between formation and breakdown.

But now closer examination of brain structure suggests that it, too, is dynamic, that it, too, is undergoing continuous change and remodeling. Indeed, it is interesting to note that from the perspective of general systems theory structure is defined as slow process, whereas function is described as rapid process (Miller, 1975). Indeed, if we look back at the anatomical studies of the ecology of brain, we find that most structures which have been examined to date, from whole brain size down to synaptic morphology, appear to be plastic. Time-lapse photography of brain tissue cultures suggests that neural tissue may be continuously active and mobile. Ultra-structural remodeling may take place continuously throughout life (Sotelo and Palay, 1971), and synapses may be formed and dissolved as shifts in demands, patterns of input, and dynamic equilibria occur. Indeed, a variety of morphological changes may be evident within minutes of stimulation (e.g., Van Harreveld and Fifkova, 1975b). Even in the myelin sheath, which was long thought to be immutable, studies suggest that some molecular components may be exchanged throughout the lifetime.

Thus the picture of brain that is emerging is truly a dynamic one at all dimensions, from the structural to the chemical. Descriptions in terms of states may be necessary, but it should be remembered that these are only still pictures, or shorthand notations, for descriptions of the processes which underlie them. No state is fixed or permanent, all undergo change: the only question is whether we look with sufficient sensitivity to recognize this.

The universality of dynamism and impermanence is not a novel discovery, but one which recapitulates that of modern physics, which in turn is a recognition of the wisdom of both ancient and modern mystics. In the middle size range of our day-to-day life, between the atomic and astronomical, and through the concretizing and stasis producing filters of our concepts and perceptual limitations, the world appears relatively fixed and certain. But if we refine our perceptual sensitivity, if we examine either the very large or the very samll, matter or mind, we find nothing solid, nothing fixed, only eternal ceaseless motion. Any object, when examined closely enough, turns out to be dynamic and alive, a process rather than an event. As Capra (1975: p.178) notes:

> Modern physics, too, has come to conceive of the universe as such a web of relations and, like Eastern mysticism, has recognized that this web is intrinsically dynamic. The dynamic aspect of matter arises in

> quantum theory as a consequence of the wave-nature of subatomic particles, and is even more essential in relativity theory, as we shall see, where the unification of space and time implies that the being of matter cannot be separated from its activity. The properties of subatomic particles can therefore only be understood in a dynamic context; in terms of movement, interaction, and transformation.

The more closely we examine anything, the more dynamic it appears. Molecules move in ceaseless Brownian motion, within the atoms the electrons race around the nuclei, and within the nuclei themselves protons and neutrons race at speeds approaching that of light.

But at this subatomic level, structures are processes not only by virtue of their rapid motion, but also because their intrinsic nature is one of changing probability patterns and interconnections of energy rather than of material substance (Capra, 1975, 1976). At the other end of the size spectrum, the same inherent findings holds true: always motion, always change. Even when we push to the farthest reaches of the known universe, we find dynamism, a universe expanding at speeds approaching that of light.

Change and impermanence are nothing new to those who have explored the farther reaches of consciousness. Indeed, these tenets of the fundamental nature of reality are central to several disciplines. For Buddhists, one of the most profound insights available to any person is that not only is everything in the outer world and in the body and mind undergoing ceaseless change, but also in the depths of our consciousness that which we assume to be ourselves, our "Self" or "I," is also in continuous flux. When one looks deeply enough, there is no solid, consistent, or permanent "I," only a ceaseless and impersonal flux of thoughts and emotions which arise into consciousness and then pass away in such rapid succession that the untrained awareness perceives an illusion of stability and continuity (Thera, 1962; Goldstein, 1976; Goleman, 1977; Walsh, 1977, 1978). As without, so within: nothing stable, nothing solid, nothing permanent, only a Heraclitean flow of such rapidity that our normal perceptual processes are inadequate to the task, and construct a reality of solidity and stasis. Indeed, one of the Buddha's major teachings was that much of the cause of human suffering lay in the impossible search for security; for what can be secure when everything changes? It is the direct experience rather than the merely intellectual understanding of this fact which constitutes the basis for the Buddhist meditation called "Insight" or "Vipassana."

During Vipassana training the meditator passes through a well-mapped sequence of degrees of perceptual sensitivity, speed, and concentration, eventually reaching levels quite beyond those currently recognized and examined by Western psychology. At these stages, not only is the arising and passing away of each thought and emotion clearly noted, but also mental processes persisting for only a minute fraction of the period during which a thought is present in awareness are clearly perceptible (Shapiro and Walsh, 1978; Walsh and Shapiro, 1978o; Walsh, 1977, 1978; Goldstein, 1976; Goleman, 1977). At this level of sensitivity, the dynamic, impermanent, and impersonal nature of all experience is inescapable, and initially causes considerable dysphoria as the meditator is forced to let go previous models of self and reality. However, it is this heightened perception and insight, and the dropping of inappropriate models, that clear the way for the change in consciousness necessary for the deepest levels of understanding and cessation of suffering. In the words of Radhakrishnan (1958: pp.367, 369):

> A wonderful philosophy of dynamism was formulated by Buddha 2,500 years ago . . . impressed with the transitoriness of objects, the ceaseless mutation and transformation of things, Buddha formulated a philosophy of change. He reduces substances, souls, monads, things; to forces, movements, sequences, and processes and adopts a dynamic conception of reality . . .

From D. T. Suzuki (1968: p.55):

> How do we come to think of things, rather than of process in this absolute flux? By shutting our eyes to the successive events. It is an artificial attitude that makes sections in the stream of change, and calls them things . . . When we shall know the truth of things, we shall realize how absurd it is for us to worship isolated products of the incessant series of transformations as though they were eternal and real. Life is no thing or state of a thing, but a continuous movement or change.
>
> Buddhists have conceived an object as an event and not as a thing or substance . . . the Buddhist conception of "things" as samskara (or sankhara), that is, as "deeds" or "events," makes it clear that Buddhists understand their experience in terms of time and movement.

Thus, deep investigation in either science or the consciousness disciplines forces the recognition of the dynamic nature of the universe. In science, this leads to a recognition of the need for not only dynamic formulation but also experimental designs which allow the recognition of process. Ignorance of, or simply ignoring, such temporal factors as biological rhythms and temporal profiles of response have often led to reduced sensitivity and validity of findings in the past (Walsh and Cummins, 1976b); the more sophisticated and sensitive the science, the more important the recognition of the dynamic nature of reality.

## Probabilism and Uncertainity

The observation that nothing is certain in life may hold across levels from the most trite to the most fundamental, and from the macroscopic to the quantal. Absolute statements may suffice for first order approximations, but more precise measurements and descriptions necessitate a statistical discussion. Behavioral and neural events can rarely be predicted with absolute certainty. Even the most apparently simple behaviors, responses, and systems tend to show unanticipated variability when examined closely enough. Even that prototype of organismic simplicity, the unicellular amoeba, displays an unpredictable pattern of behavioral responses to a simple pinprick stimulus. A rat trained to asymptotic levels in a simple, two-choice maze or discrimination apparatus will occasionally select the nonrewarded choice. Single neurons, and that most complex of all neural structures, the cerebral cortex, can be viewed as statistical integrators which receive input from a wide variety of sources and respond with statistically appropriate discharge parterns (Sholl, 1956). The nature of the experimental designs (large groups) employed to examine the ecology of brain attests to the probabilistic nature of effects. It is an unusual experiment which does not find one or more littermate pairs in which the direction of the environmental effect is reversed, compared with the statistical mean of the group.

Taken individually, these examples appear relatively trite, but they point to something of considerable significance. In any complex, highly over-determined system, it is to be expected that it will be impossible to control more than a small fraction of the determinants. Although a major portion of the variance may be controlled, the uncontrolled variables will superimpose a degree of unpredictability. Therefore, precise examination and description will necessitate a statistical approach.

Thus, although initial examinations may show apparently nonprobabilistic behavior, more precise observation will reveal the concomitant unpredictable component. And with increasing sensitivity of observation will come increasing appreciation of the probabilistic nature of the behavior of the system.

This probabilistic nature of experimental results has been detected at the most fundamental levels of atomic physics, where it seems to be an inherent property of both measurement procedures and the particles which they examine. Indeed, subatomic particles cannot even be said with certainly to exist, but rather must be described as showing "tendencies" to exist or to occur.

Furthermore, at this level it seems that there are inherent limitations in our ability to measure pairs of quantal conjugated variables (Heisenberg, 1958). Thus, for example, the very measurement procedure which allows us to know the momentum of an electron prevents our measuring its position, and vice versa. This limitation is known as Heisenberg's Principle and is not restricted to this situation alone, but includes other examples of conjugated variables, such as the time an atomic event takes and the energy involved. The precision and certainty of the measurement of one variable limits to a corresponding degree the precision and variability of the other. In addition to noting the inherent limitations of measurement, we might ask to what extent the universe must be seen as probabilistic as a function of the probabilistic nature of brain function.

It was partially in response to these limitations that Bohr (1934) introduced the Complementarity Principle. Herein he advanced the idea that the image of a subatomic unit as a particle and the image of it as a wave were two complementary pictures of reality, each a partially correct and limited picture of reality which in no way negated the reality of the other. Both perspectives are essential for the fullest description of reality, and both may be applied within the limitations imposed by the uncertainty principle.

As Capra notes, the complementarity notion has become an essential part of the world view of modern physics, and Bohr has often suggested its usefulness in other areas. For 2,500 years, Chinese Taoistic thought has viewed opposite concepts as polar and complementary. This was archetypally represented by the opposites of yin and yang, which were seen as representing fundamentally opposite forces operating throughout nature which through their interplay determined the nature and behavior of man and universe (Needham, 1956).

The importance of the Complementarity Principle is well described by Capra (1975: pp.144-145).

> The fundamental importance of the uncertainty principle is that it expresses the limitations of our classical concepts in a precise mathematical form. As described previously, the subatomic world appears as a web of relations between the various parts of a unified whole. Our classical notions, derived from our ordinary macroscopic appearance, are not fully adequate to describe this world. To begin with, the concept of a distinct physical entity,

> like a particle, is an idealization which has no functional significance. It can only be defined in terms of its connections to the whole, and these connections are of a statistical nature-probabilities rather than certainties. When we describe the properties of such an entity in terms of classical concepts-like position, energy, momentum, etc. - we find that there are pairs of concepts which are interrelated and cannot be defined simultaneously in a precise way. The more we impose one concept on the physical 'objects', the more the other concept becomes uncertain, and the precise relation between the two is given by the uncertainty principle.

Uncertainty is known to the advanced meditator at the very deepest levels and is called by many names, including "great doubt" and "profound unknowing". The profound experience of ceaseless change and impermanence forces a deep recognition of the unpredictability and uncertainty of all things. Growth leads to a deeper and deeper sense of the mystery and unfathomability of all things (Rajneesh, 1975a, 1975b). As noted previously, the attachment to, and search for, certainty in an inherently uncertain universe is held to be one of the major sources of human suffering. Likewise, attachment to any extreme, any pole, with consequent avoidance or neglect of its complementary opposite is seen as resulting in disequilibrium and disharmony with the flow and the Tao. Thus, the person of wisdom therefore seeks the ability to encompass and flow smoothly between extremes without becoming stuck in either (Lao Tsu, 1972), and to experience the universe and his or her consciousness as fully as possible ("to reach the totality of his being" [Castaneda, 1971]), without demanding that this totality should be other than it is, mysterious, unfathomable and uncertain.

## Causality and the Boot Strap Universe

One of the major guiding principles and aims of science has been to describe phenomena in terms of the smallest feasible number of explainer variables or mechanisms. Therefore, typical experimental and conceptual approaches usually consider only the major independent variables and ignore others. There is also a strong tendency towards breaking structures down into their constituents. This approach can be traced all the way back to the atomistic philosophy of the Greeks, which held that the workings of the universe were to be understood in terms of the actions and interactions of fundamental indivisible units called atoms, and is a type of approach that still tends to guide our search for causes and mechanisms. Thus, for example, in the attempt to explain the neural effects of environmental complexity or the nature of learning, there is a tendency to look for the underlying mechanism, the single structure or process which will account for these phenomena (e.g., the molecule, the chemical process, the neurophysiological mechanism, the subcellular component, the synaptic modification).

However, such an approach, though perhaps useful as an initial step, may be fundamentally inappropriate. With increased sophistication and sensitivity there arises a greater recognition of the need to take into consideration increasing numbers of contributing independent variables, and increasing numbers and complexity of interconnections and interactive effects. The ultimate recognition is of the total interconnectedness and interdependence of all structure-processes: a recognition beautifully encapsulated in the proverb, "If you cut a blade of grass you shake the universe."

As an example, let's take a closer look at the determinants of the cerebral response to environmental complexity. The original assumption was that compared to stimulus deprivation, the complex environment would elicit an asymptotically larger cortex with increasing duration of exposure. So far, so simple. However, in recent years, as other independent variables have been examined, it has been found that the effect represents the result of a multi-way interaction in which practically all independent variables examined to date play an appreciable part. Thus, for example, the magnitude and even direction of the cerebral response to environment depends on such stimulus characteristics as its nature, intensity, and duration and when during biological cycles it is applied. The parameter under investigation and the area in which it is examined also play a role, as do a variety of characteristics of the experimental subject. Thus, for example, the subject's age, sex, genetic background, experiential background, nutrition, and hormonal status all play a part. It should be noted that these effects are not necessarily simply additive, but rather may be interactive so that the final outcome cannot be predicted in a simple additive fashion. This is even more apparent when we come to examine behavior, where interactive effects may be so multifactorial and sensitive that it is questionable to what extent the results are generalizable outside the particular laboratory in which they were performed (Henderson, 1969; Walsh and Cummins, 1976b).

To complicate things still further, we must recognize that causality is not a simple one-way affair, and that the prototype of traditional causality models, the stimulus-response model, may well have a limited useful lifespan. Even in that archetype of stimulus-response paradigms, behavioral modification, there have recently emerged more complex models of reciprocal determinism (Bandura, 1978). This is probably a useful concept, though, of course, it still remains a gross oversimplification, since in view of the fundamental interconnectedness of all entities nothing short of a multiway reciprocal determinism, mutual causality, or "omni-determinism" will suffice.

From this vantage point of heightened recognition of the complexity of causality, let's take another look at the process of neural responses to sensory environments. Formerly we considered this process in terms of a simple stimulus-response model with a variety of mechanisms mediating the effect of environment on brain. However, this is obviously a gross oversimplification, and we can make it somewhat more realistic by including two additional factors: firstly, the reciprocal action of the organism on the environment, and secondly, the role of consciousness.

Obviously, an organism's behavior will modify the environment so as to create a reciprocal determinism between them. In one sense, the behavior can be seen as part of the environment, since it provides a source of stimulation (reafferent) which may be especially potent in its effects on brain.

Preceding portions of this chapter should suggest the appropriateness of including consciousness in the model, for we have seen that what is preceived is as much a function of consciousness as it is of whatever may be "out there." Furthermore, as we shall see, it may even be that consciousness modifies brain.

If we now examine just the interactions between these three systems, brain, environment, and consciousness, we can develop a more comprehensive, reciprocally deterministic model with each component varying as a function of the others.

1. Brain $\alpha$ F (environment)
   i.e., Environment affects brain structure and function
2. Environment $\alpha$ F (brain)
   i.e., The environment may be modified by behavior
3. Consciousness $\alpha$ F (brain)
   i.e., Brain activity determines consciousness
4. Environment $\alpha$ F (consciousness)
5. Brain $\alpha$ F (consciousness)

Equations 1, 2 and 3 are straightforward. Equation 4 states that environment as it is perceived is a function of consciousness, a point which has been discussed in detail previously.

The fifth equation states that brain is a function of consciousness, and here there are two possible major mechanisms. Sperry (1977) has advanced the hypothesis that brain structure and function may be modifiable by consciousness. Using general systems theory, he considers consciousness to be a so-called emergent function of brain, a function or property of the gestalt which is more than can be accounted for by the sum of the component parts, and he points out that in other systems such emergent functions often control subcomponents.

A second possible mechanism consists of "reactivity". When a behavior is observed closely by the individual, that behavior usually changes without conscious effort (Thoresen and Mahoney, 1974). Such behavior changes presumably are accompanied by concomitment neural effects.

Of course, even this expanded model is simplistic, since each of the systems examined above consists of interacting and interdetermining subsystems, and ultimately we are forced to recognize the inherent limitations and inadequacy of all models. This omni-determinism and the limited applicability of traditional models of causality are problems which the consciousness disciplines and physics have had to confront, and from each of them have emerged parallel pictures which transcend our usual models of causation.

In physics the "boot strap hypothesis" takes as its basic premise the idea that nature cannot be reduced to fundamental entities, be they particles, forces, fields, or mechanisms. Rather, nature is to be understood entirely in terms of interdependent self-consistency, in which all its component parts determine, and are consistent with, themselves and each other. The interconnected web of events which is the universe, is not to be understood in terms of the fundamental properties of individual units, but rather in terms of the whole and the totality of their interactions and interrelations. Such a view represents the culmination of the evolution away from the Newtonian mechanistic world view. There are no fundamental or absolute entities of any kind awaiting discovery, only a mutual causality or omnideterminism. All that we discover are co-creations of the human mind: properties of our conceptual maps as much as of the world they describe.

The recent evidence in favor of the existence of quarks calls the precise formulations of the original bootstrap theory (Chew, 1968) into question. However the general principles of holism and interdependence remain and are also components of other recent hypothesized world views in physics (e.g. Bohm, 1978).

Mystics have long held a similar understanding of the nature of reality and have clearly been aware of the limits of an atomistic approach, of the interconnected, interpenetrating nature of things, and of the impossibility of describing them in terms of traditional causality models.

> The least act, such as eating or scratching an arm,
> is not at all simple. It is merely a visible moment in a

> network of causes and effects reaching forward into Unknowingness and back into an infinity of Silence, where individual consciousness cannot even enter. There is truly nothing to know, nothing that can be known.
>
> The physical world is an infinity of movement, of Time-Existence. But simultaneously it is an infinity of Silence and Voidness. Each object is thus transparent. Everything has its own special inner character, its own Karma or 'life in time,' but at the same time there is no place where there is emptiness, where one object does not flow into another (Kapleau, 1967: p.268).

As regards atomism, Ashvaghosha (1900: p.104) remarked:

> When we divide some gross matter, we can reduce it to atoms. But as the atom will also be subject to further division, all forms of material existence, whether gross or fine, are nothing but the shadow of particularization and we cannot ascribe any degree of reality to them.

Perhaps the most exquisite metaphor for the interpenetrating network of reality is that of Indras' net, which is a vast network of scintillating gems which hangs over the palace of the god Indra and pictures interconnection.

> In the heaven of Indra, there is said to be a network of pearls, so arranged that if you look at one you see all the others reflected in it. In the same way each object in the world is not merely itself but involves every other object and in fact is everything else. 'In every particle of dust there are present Buddhas without number' (Eliot, 1969: pp.109-110).

Such a view transcends the capacity of our everyday concepts. "In the Absolute there is neither time, space, nor causation" (Vivekananda, 1972: p.109).

To this ultimate finality
no law or description applies

Sengtsan, 1975

In the words of Rajneesh (1975a: pp.29-31)

Existence is -
There is no why about it . . .
Every answer leads you into a deeper mystery.

Looking back from this perspective we now ask, what, then, are the mechanisms mediating learning and environmental complexity? Strictly speaking, the mechanism is the total change - everything is involved in and constitutes the mechanism, and no one subsystem or its alterations can be isolated, set apart, and looked to as "it." According to data or predilection, we can abstract out any subset of processes and suggest that they constitute the underlying mechanism(s), but each such abstraction will represent only one perspective of the totality which ultimately lies beyond any single perspective.

Such a recognition is a far cry from science as it is usually practiced, and necessitates the exploration of how to incorporate this perspective into both research and theory. Since we can perform only finite and limited experiments, and abstract correspondingly limited perspectives and theories, obviously we are faced with a discrepancy between the true nature of reality and the pictures our experiments and conceptual abilities can provide.

It may be that the best we can do is to choose the optimal experimental strategies from within the context of an awareness of their limitations and of the indivisible, holistic nature of that which we are seeking to examine. As Globus and Franklin (1976) have suggested, we need a dynamic epistemology in which we employ different and complementary epistemological tools in recognition of the arbitrariness and incompleteness of the picture provided by any one approach. Sophisticated descriptions will rarely be in terms of a single level, factor, or modality. Rather, they need to be multifactorial and multilevel, with the recognition that one level is not necessarily more fundamental than another, and that all factors and levels are interconnected and interdependent. The picture we obtain is but an arbitrary, finite, fragmented, relativistically real, co-created view of an infinite, indivisible, acausal whole. Such a context reduces the risk of confusing our conceptual map knowledge with the reality it attempts to describe.

In summary then, a number of findings emerging from the study of the effects of the environment on brain appear to display parallels to certain claims of modern physics and the consciousness disciplines. The view of nature emerging from these parallels appears to represent a function of heightened perceptual and experimental sensitivity and as such it is hypothesized that as other branches of science continue to develop they too will recognize these same parallels. Only the future can tell us whether this is correct.

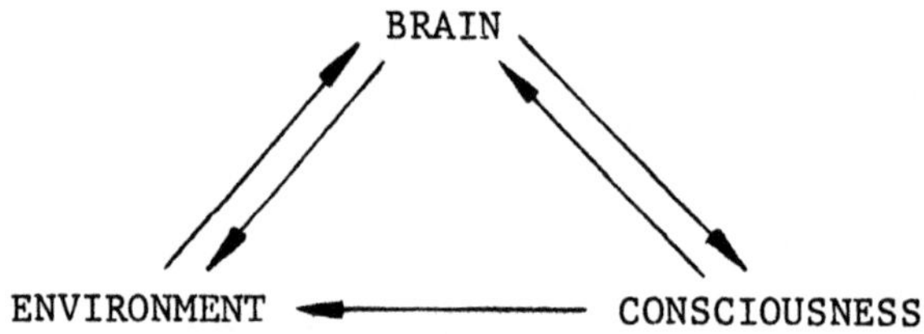


Fig. 1. Multiway reciprocal deterministic model of the interaction of brain, environment, and consciousness.

Fig. 2. The ultimate outcome of environmentally enhanced brain development.

# CHAPTER 10

# An Overview and Glimpse at the Future of the Ecology of the Brain

Less than a quarter of a century has passed since the first clear demonstration of the ecological nature of the brain. Within that time, however, the field has grown remarkably, both in terms of its data, and its overlap and convergence with other fields, and in terms of recognition of its broader implications. The ecology of the brain clearly represents a nodal point for the interaction of a variety of research fields and of much new thinking.

The emerging picture is one of a dynamic brain responding to its sensory environment at all levels, from the gross anatomical to the ultrastructural. The changes observed to date are consistent with an adaptational model of neural expansion and differentiation or contraction, occurring in response to, and allowing optimal processing of, sensory input. Whether such responses are always optimal or whether sometimes they are maladaptive, and if so under what conditions, is unknown.

These responses have obvious teleological significance and explode a variety of neuromythologies. Foremost of these, of course, is the old idea of the immutability of brain, and its lesser version, the notion that if there is any brain response it will occur only in young, developing organisms. It is clear that the structure and function of the brain can no longer be considered in isolation from its environment. Nor can the effects of other brain modifying agents, such as drugs, be considered in isolation or even in simple additive terms. Rather, it is apparent that the necessary perspective is an ecological one which recognizes both the sensitivity of the brain to a variety of factors and the interactive nature of these agents in their determination of any outcome.

The range of parameters found to be responsive to the sensory environment continues to increase. Indeed, it now seems appropriate to stop wondering whether a response will occur and to assume it will; the question then becomes how large the response will be, and whether we will be able to detect it experimentally. As in many other areas of neuroscience, the specificity of brain responses as regards region, cell type, and temporal patterning is becoming apparent, and the error of treating the brain as homogenous is clearly evident.

If we remember that it is less than a quarter century since the brain was widely assumed to be immutable, then we are left to wonder just how large its responses can be and what the limits of plasticity truly are. As our knowledge expands and becomes more precise concerning the nature of the factors which enhance responsivity, we may yet look back with amazement a quarter century hence at our own limited appreciation of the brain's potential.

On the stimulus side, presumably there will be a greater effort at component analysis to determine which stimuli and combinations of stimuli are most effective, and what the precise characteristics which constitute efficacy are. At the present time, stimuli are usually categorized in terms of distal variables, such as environmental complexity and social grouping, which are actually complex, multidimensional stimulus situations. With increased sophistication there will probably be a tendency towards a more precise and proximal description of variables. For social parameters, this will entail a closer examination of the number and nature of interactions between individuals, and might even lead towards a field of socio-neurobiology. Behavior, including the social, will probably be examined more closely for its role as an intermediary variable mediating the effects of more distal ones.

The increasing recognition of responses as processes rather than as static end points suggests the necessity of determining temporal response profiles. These, together with studies of dynamic equilibria and chemical turnover, should lead to a greater appreciation of the dynamic nature of brain.

Now that the ecological nature of the brain is clearly evident, a whole new field for investigation of interaction between sensory and nonsensory factors opens. There is already evidence that the effects of chemicals, hormones, nutrition, and biological rhythms cannot be determined in isolation, but rather as they interact with the effects of environment. This recognition holds great potential for optimizing the effectiveness of both environmental and nonenvironmental factors, such as for therapeutic goals.

It also points clearly to the necessity for reassessing research fields which have been unaware of the environmental sensitivity of the brain. For example, it is questionable whether more than 5% of the studies of the effects of malnutrition can be accepted unequivocally, since it is now apparent that many of the experimental paradigms employed to induce malnourishment vary along stimulus dimensions such as litter size or maternal-infant interaction. These factors are in and of themselves capable of eliciting significant changes in brain and behavior.

This recognition may be particularly important for understanding brain damage, since it is now clear that the environment may play a role in both the etiology and the therapy of dysfunction, and that the damaged brain may be especially sensitive to variations in stimulation. At least some of the disabling effects of damaging agents may be mediated by sensory factors, such as reduced attention to sensory input. On the other hand, it is also apparent that sensory environments may hold therapeutic potential considerably greater than has been recognized to date, and that optimal combinations of sensory and nonsensory factors administered in appropriate amounts and at appropriate times during the recovery period may be effective where either alone may fail. Certain stimulus deprivation paradigms, such as unilateral eye closure, may also provide useful models for studying clinical syndromes like neurological speech deficits, whose effects partially represent the result of contralateral inhibition. At the very least, it is important that clinicians become more aware of the sensitivity of the damaged brain to its environment.

There are also broad implications for a variety of other applied fields. For example, many of our child rearing practices must be questioned. These include the practice of separating infants and mothers at birth in spite of the fact that it has been clearly demonstrated in both humans and animals that this may produce subsequent deficits in maternal behavior. The practice of placing sick neonates in stimulus depriving isolette cots must also be questioned. Similarly, one wonders about the deleterious effects of solitary confinement on prisoners, and the effects of social and physical stimulus deprivation on the handicapped and many of our geriatric population. Concerning this last group, we must also note that it becomes all the more important in view of studies demonstrating the unexpected plasticity and responsiveness of the geriatric brain. There is an urgent need for further studies of the efficiency of enrichment programs for both infants and the aged. Perhaps the intellectual decline associated with old age can be diminished or even reversed with appropriate stimulation. Similarly, better planned studies may yet prove extremely helpful in ameliorating the so-called cultural deprivation syndrome. In these days of such concern over genetic versus environmental determination of intelligence, it is well to remember that we still have not the faintest clue about how much intelligence could be increased in various population groups by an "optimal environment."

In terms of more general implications, it is apparent that studies of brain responsivity should now include controls for sensory input. It is unfortunately necessary to include a further reminder of the importance and necessity of performing studies blind, that is, without knowing whether subjects are derived from experimental or control groups, in order to minimize subjective bias.

The recognition of the ecological nature of brain is in accord with studies in a variety of fields which have recently begun to emphasize the importance of context, in both determining and understanding outcome. Studies of units in isolation may be necessary at times, but may also entail a greater loss in understanding than has previously been recognized. The implications of this ecological context for brain are far reaching indeed, extending beyond the neurosciences towards convergences with other disciplines, both scientific and nonscientific, and even to the boundaries of science and the intellect. The inherent inseparability of the brain from the universe which it is designed to know leads beyond traditional dichotomous paradigms, demands radical reconceptualizations, and even points to the limits of science and of concepts themselves.

Ultimately we are forced to the level of paradox, as we find ourselves to be the co-creaters of the reality we seek to describe in our movement towards an ecology of the brain.

> We have only just begun, with the real questions and answers still ahead.
>
> (Diamond, 1976: p.239)

# References

It is customary to ornament every scientific work with a bibliography, a list of earlier books on the same subject. The aim may be to test the reader's competence by laying out for him an awe-inspiring course of reading; it may be to prove the author's competence by observing the mountain of dross he has sifted to win one nugget of truth.
(Peter and Hull, 1969, p. 79)


Adler, M.W., Bendotti, C., Ghezzi, D., Samanin, R., and Valzelli, L. 1975a. Dependence to morphine in differentially housed rats. Psychopharmacologia 41, 15-18.

Adler, M.W., Mauron, C., Samanin, R., and Valzelli, L. 1975b. Morphine analgesia in grouped and isolated rats. Psychopharmacologia 41, 11-14.

Agrawal, H.C., Fox, M.W., and Himwich, W.A. 1967. Neurochemical and behavioral effects of isolation-rearing in the dog. Life Sci. 6, 71-78.

Ainsworth, M.D. 1973. The development of infant mother attachment, In Review of Child Development Research, Vol. 3. Caldwell B.N. and Riccuti, H.N. eds. University of Chicago Press, Chicago. 1-94.

Allee, W.C. 1942. Group organization among vertebrates. Science 95, 289-293.

Altman, J. 1966. Autoradiographic examination of behaviorally induced changes in the protein and nucleic acid metabolism of the brain, Macromolecules and Behavior. Gaito, J. ed. Appleton-Centure-Crofts, New York.

Altman, J. 1967. Postnatal growth and differentiation of the mammalian brain, with implications for morphological theory of memory, in The Neurosciences, A Study Program. Quarton, G., ed., Rockefeller Press, New York, 723-724.

Altman, J. 1970. Postnatal neuronogenesis and the problem of neural plasticity, in Developmental Neurobiology Himwich, W.A. ed. Thomas, Springfield, 194-237.

Altman, J. Anderson, W.J. and Wright, K. 1968a. Gross morphological consequences of irradiation of cerebellum in infant rats with repeated doses of low level X-ray. Exp. Neurol. 21, 69-91.

Altman, J. Anderson, W., and Wright, K. 1968b. Differiential radiosensitivity of stationary and migratory primitive cells in the brains of infant rats. Exp. Neurol. 22, 52-74.

Altman, J., and Das, G.D. 1964. Autoradiographic examination of the effects of enriched environment on the rate of glial multiplication in the adult rat brain. Nature (London) 204, 1161-1163.

Altman, J., Wallace, R., Anderson, W., and Das G. 1968c. Behaviorally induced changes in length of cerebrum in rats. Dev. Psychobiol. 1, 112-117.

Appel, S.H., Davis, W., and Scott, S. 1967. Brain polysomes: response to environmental stimulation. Science 157, 836-838.

Ashvaghosha. 1900. The Awakening of Faith. Suzuki, D.T., trans. Open Court, Chicago, Illinois.

Axelrod, J. 1974. Science 184, 1341.

Bandura, A. 1969. Principles of Behavior Modification. Holt Rinehart, and Winston, New York.

Bandura, A. 1977. Social Learning Theory. Prentice-Hall, Englewood Cliffs, New Jersey.

Bandura, A. 1978. The self system in reciprocal determinism. Am. Psychol. 33, 344-358.

Barker, R. 1968. Ecological Psychology. Stanford University Press. Stanford, California.

Barnett, S.A., and Walker, K.Z. 1974. Early stimulation, parental behavior, and the temperature of infant mice. Dev. Psychol. 7, 563-577.

Barondes, S.H. 1967. Neurosci. Res. Program Bull. 5, 307.

Batkin, S., Groth, H., Watson, J., and Ansberry, M. 1970. Effects of auditory deprivation on the development of auditory sensitivity in albino rats. Electroencephalagr. Clin. Neurophysiol. 28, 351-359.

Baumel, I., DeFeo, J.J., and Lal, H. 1970. Alterations in brain sensitivity and barbiturate metabolism unrelated to aggression in socially deprived mice. Psychopharmcologia 18, 610-619.

Baumel, I., DeFeo, J.J., and Lal, H. 1969. Decreased potency of CNS depressants after prolonged social isolation in mice. Psychopharmacologia 15, 153-158.

Beck, E.C., Dustman, R., and Sakai, M. 1969. Electrophysiological correlates of selective attention, in Attention in Neurophysiology, Evans, C., and Mulholland, T., eds., Butterworths, London.

Bennett, E.L. 1976. Cerebral effects of differential experience and training, in Neural Mechanisms of Learning and Memory. Rosenzweig, M.R. and Bennett, E.L. eds., M.I.T. Press, Cambridge, Mass. 279-287.

Bennett, E.L., Diamond, M.C., Krech, D., and Rosenzweig, M.R. 1964. Chemical and anatomical plasticity of brain. Science 146, 610-619.

Bennett, E.L., and Rosenzweig, M.R. 1968. Brain chemistry and anatomy: Implications for theories of learning and memory, in Mind As a Tissue, Rupp, C. ed. Hoeber Medical Division, Harper and Row, New York, 63-86.

Bennett, E.L., and Rosenzweig, M.R. 1971. Chemical alterations produced in brain by environment and training, in Handbook of Neurochemistry, Lajtha, A. ed., Plenum Press, New York, 173-201.

Bennett, E.L., Rosenzweig, M.R., and Chang-Wu, S.Y. 1973. Excitant and depressant drugs modulate effects of environment on brain weight and cholinesterases. Psychopharmacologia 33, 309-328.

Bennett, E.L., Rosenzweig, M.R., and Diamond, M.C. 1969. Rat brain: effects of environmental enrichment on wet and dry weights. Science 163, 825-826.

Bennett, E.L., Rosenzweig, M.R., Diamond, M.C., Morimoto, H., and Herbert, M. 1974. Effects of successive environments on brain measures. Physiol. Behav. 12, 621-632.

Berger, H. 1900. Experimental anatomuscle studien uher die durch den mangel optiocher Reize veranlassten Entruichlungshemmungen in occipitallappen des Kumdes and der Katze. Arch. Psychiat. 33, 521-567 as cited by Riesen, 1966.

Berlyne, D.E. 1960. Conflict, Arousal and Curosity. McGraw-Hill, New York.

Bernstein, L. 1971. The reversibility of learning deficits in early environmentally restricted rats. J. Psychosom. Res. 15, 229-232.

Bernstein, L. 1972. The reversibility of learning deficits in early environmentally restricted rats as a function of amount of experience in later life. J. Psychosom. Res. 16, 71-73.

Bernstein, M.E., and Bernstein, J. 1973. Regeneration of axons and synaptic complex formation rostral to the site of hemisection in the spinal cord of the monkey. Int. J. Neurosci. 5, 15-26.

Bertalanffy, L. Von. 1969. General Systems Theory. Braziller, New York.

Beteleva, T.G. and Novikova, L.A. 1961. Electrical activity in various cortical regions and in the reticular formation after elimination of the olfactory analyzer. Pavlov J. Higher Nerv. Act. 11, 547-555.

Bevan, W., Jr., Bloom, W.L., and Lewis, G.T. 1951. Levels of aggressiveness in normal and amino acid-deficient albino rats. Physiol. Zool. 24, 231-237.

Bewsher, P.D., Hillman, C.C., and Ashmore, J. 1966. Studies on the hypoglycemic effect of d-amphetamine in aggregated mice. Biochem. Pharmacol. 15, 2079-2085.

Beynam, L.M. 1978. The emergent paradigm in science. Re-Vision 1, 56-72.

Birren, J.E., and Hess, R.D. 1968. Influences of biological, psychological, and social deprivations upon learning, and performance, in Perspectives on Human Deprivation: Biological, Psychological and Sociological. U.S. Department of HEW, Washington, D.C. 89-193.

Bjorklund, A., and Stenevi, V. 1972. Nerve growth factor: stimulation of regenerative growth of central noradrenergic neurons. Science 175, 1251-1253.

Bogoch, S. 1962. In Cerebral Sphingolipidoses. Aronson, S.M, and Volk, B.W., eds. Academic Press, New York, 249.

Bogoch, S. 1965. Brain proteins in learning: Findings in the experiments in progress discussed at the NRP work session, June 27-29, 1965. Neurosci. Res. Program Bull. 3, 38-41.

Bogoch, S. 1968. The Biochemistry of Memory. Oxford University Press, Oxford, England.

Bogoch, S. 1972. Brain glycoprotein 10B: Further evidence of the sign-post role of brain glycoproteins in cell recognition, its change in brain tumor, and the presence of a "distance factor," in Functional and Structural Proteins of the Nervous System, Davidson, A.N., Mandel, P., and Morgan, I.G. eds. Plenum Press, New York 39-52.

Bogoch, S. 1973. Brain glycoproteins and learning: New Studies supporting the "sign-post" theory. Current Biochemical Approaches to Learning and Memory. Essman, W.B., and Nakajima, S. eds. Spectrum-Halstead Press, New York, 147-157.

Bogoch, S., and Belval, P.C. 1966. In Inborn Disorders of Sphingolipid Metabolism. Aronson, S.M., and Bolk, B.W. eds. Pergamon Press, New York, 273.

Bohm, D. 1978. The enfolding-unfolding universe. ReVision 1(3/4), 24-51.

Bohm, D., and Hiley, B. 1975. On the intuitive understanding of non-locality as implied by quantum theory. Foundations of Physics 5, 96.

Bohr, N. 1934. Atomic Physics and the Description of Nature. Cambridge Univ. Press, Cambridge, England.

Bohr, N. 1943. Atomic Physics and the Description of Nature, Cambridge Univ. Press, Cambridge, England.

Bohr, N. 1958. Atomic Physics and Human Knowledge. John Wesley, New York.

Bondy, S.C., and Roberts, S. 1969. Developmental and regional variations in ribonucleic acid synthesis on cerebral chromation. Biochem. J. 115, 341-349.

Bonner, J., Chalhley, G., Darmus, M., Farmbrough, D. 1968. In Methods in Enzymology, Grossman, L., and Moldave, K., eds. Vol. 12B Academic Press, New York, 3-65.

Bourgault, P.C., Karczmar, A.G., and Scudder, C.L. 1963. Contrasting behavioral, pharmacological, neurophysiological, and biochemical profiles of C57, B1/6 and SCI strains of mice. Life Sci. 8, 533-553.

Brain, P.F., and Nowell, N.W. 1969. The effects of isolation as opposed to grouping on adrenal and gonadal functioning in male and female mice. J. Endocrinol. 46, XVI-XVII.

Braun, J., Meyer, P., and Meyer, D. 1966. Sparing of a brightness habit in rats following visual deterioration. J. Comp. Physiol. Psych. 61, 79-82.

Brenner, C. 1974. An Elementary Textbook of Psychoanalysis. (2nd ed.) Anchor Books, New York.

Bresler, D., and Bitterman, M. 1969. Learning in fish with transplanted brain tissue. Science 163, 590-592.

Breuer, J., and Freud, S. 1965. Studies on hysteria. In Standard Edition of the Complete Psychological Works of Sigmund Freud, Vol. 2. Strachey, J. ed. Hogarth Press, London, England (Originally published, 1892).

Brizee, K.R., and Jacobs, L.S. 1959. The glia: neurone index in the submolecular layer of the motor cortex of the cat. Anat. Rec. 134, 97-105.

Bross, M., and Zubek, J.P. 1975. Effect of auditory deprivation on visual resolving power. Can. J. Psychol. 29, 340-347.

Brown, C.P., and King, M.G. 1971. Developmental environment variables important for later learning and changes in cholinergic activity. Dev. Psychobiol. 4, 275-286.

Brown, G.S. 1972. Laws of Form. Julian Press, New York.

Brown, R.T. 1968. Early experience and problem-solving ability. J. Physiol. Psychol. 65, 433-440.

Bruner, J.S. 1969. Phenotypic and genotypic variation, early stimulation and cognitive development, in Stimulation in Early Infancy, Ambrose, A. ed., Academic Press, New York, 265.

Brunngraber, E.G. 1969. The possible role of glycoproteins in neural function. Perspect. Biol. Med. 12, 467-470.

Brunswick, E. 1957. Scope and aspects of the cognitive problem, in Cognition: The Colorado Symposium, Jessor, R., and Hammond, K. eds. Harvard University Press, Cambridge, Mass.

Bryan, R.N., 1965. Brain RNA Metabolism and Differential Experience. Doctoral Dissertation, Univ. of Utah, Salt Lake City, Utah.

Cannon, W.B., and Rosenbleuth, A. 1949. The Supersensitivity of Denervated Structures, a Law of Denervation. The Macmillan Co., New York.

Capra, F. 1975. The Tao of Physics. Shambhala, London, England and Berkeley, California.

Capra, F. 1976. Modern physics and Eastern Mysticism. J. Transpers. Psychol. 8, 20-40.

Carmichael, L. 1926. The development of behavior in animals experimentally removed from the influence of external stimulation. Psychol. Rev. 34, 34-47.

Casler, L. 1965. The effects of extra tactile stimulation in a group of institutionalized infants. Genet. Psychol. Monogr. 71, 127-135.

Castaneda, C. 1971. A Separate Reality: Further Conversations with Don Juan. Simon and Schuster, New York.

Castro, A.J. 1975. Ipsilateral corticospinal projections after large lesions of the cerebral hemisphere in neonatal rats. Exp. Neurol. 46, 1-8.

Chance, M.R.A. 1946. Aggregation as a factor influencing the toxicity of sympathomimetic amines in mice. J. Pharmacol. Exp. Ther. 87, 214-219.

Chance, M.R.A. 1947. Factors influencing the toxicity of sympathomimetic amines in solitary mice. J. Pharmacol. Exp. Ther. 87, 289-296.

Changeux, J.R., and Danchin, A. 1977. Biochemical models for the selective stabilization of developing synapses, in Synapses, Cottrell, G.A. and Usherwood, P.M. eds. Academic Press, New York.

Chapman, J. 1966. The early symptoms of schizophrenia. Brit. J. Psychiat. 112, 225-251.

Cheney, D.L., Gubler, C.J., and Jaussi, A.W. 1969. Production of acetylcholine in rat brain following thiamine deprivation and treatment with thiamine antagonists. J. Neurochem. 16, 1283-1291.

Chew, G.F. 1968. "Bootstrap": A scientific idea? Science 161, 762-765.

Chow, K.L., and Stewart, D.L. 1972. Reversal of structural and functional effects of long-term visual deprivation. Exp. Neurol. 34, 409-433.

Christian, J.J. 1955. Effects of population size on adrenal glands and reproductive organs of male mice in populations of fixed size. Am. J. Physiol. 182, 292-300.

Clark, W.C., Blackman, H.J., and Preston, J.E. 1967. Certain factors in aggregated d-amphetamine toxicity. Arch. Int. Pharmacodyn. 170, 350-363.

Clarke-Stewart, K.A. 1974. Interactions between mothers and their young children. Characteristics and consequences. Monogr. Soc. Res. Child Dev. 38, (6-7, Serial No. 153).

Clendinnen, B.G., and Eayrs, J.T., 1961. The anatomical and physiological effects of prenatally administered somatotrophin on cerebral development in rats. J. Endocrinol. 22, 183-194.

Conner, R.L., Stolk, J.M., Barchas, J.D., and Levine, S. 1970. Parachlorophenylalanine and habituation to repetitive auditory startle stimuli in rats. Physiol. Behav. 5, 1215-1219.

Consolo, S., Garattini, S., and Valzelli, L. 1965a. Amphetamine toxicity in aggressive mice. J. Pharmacol. Exp. Ther. 17, 53-54.

Consolo, S., Garattini, S., and Valzelli, L. 1965b. Sensitivity of aggressive mice to centrally acting drugs. J. Pharm Pharmacol. 17, 594-595.

Consolo, S., and Valzelli, L. 1970. Brain choline acetylase and monoamine oxidase activity in normal and aggressive mice. Eur. J. Pharmacol. 13, 129-130.

Cooley, W., and Lohnes, P. 1971. Multivariate Data Analysis. John Wiley and Sons, New York.

Cragg, B.G. 1967. Changes in visual cortex on first exposure of rats to light: effect on synaptic dimensions. Nature 215, 251-253.

Cragg, B.G. 1974. Plasticity of synapses. Brit. Med. Bull. 30, 141-145.

Cragg, B.G. 1975. The development of synapses in the visual system of the cat. J. Comp. Neurol. 160, 147-166.

Craik, K. 1970. Environmental psychology. in New Directions in Psychology, Vol. 4. Holt, Rinehart and Winston, New York.

Cummins, R.A. 1978. Personal communication.

Cummins, R.A., Berwich, M., Budtz-Olsen, O.E., Walsh, R.N., and Howard, M.J. 1975. Synaptic changes in differentially reared rats. Proc. Aust. Physiol. Pharm. Soc. 6, 41.

Cummins, R.A., Carlyon, T., and Walsh, R. 1978. Drug modulated behavioral responses to environmental enrichment. Psychopharmacologia 58, 197-199.

Cummins, R.A., Livesey, P.J., Evans, J., and Walsh, R.N. 1977a. A developmental theory of environmental enrichment. Science 197, 692-694.

Cummins, R., and Walsh, R. 1978. Unpublished data.

Cummins, R.A., Walsh, R.N., Budtz-Olsen, O.E., and Cummins, M.J. 1972. New brains for old rats? Proc. Aust. Soc. Med. Res. 3, 512.

Cummins, R.A., Walsh, R.N., Budtz-Olsen, O.E., Reidel, J.C., Fry, P.C.H., and Hurley, M. 1971. Environmentally induced changes in learning ability and brain size in rats. Proc. Aust. Soc. Med. Res. 2, 477.

Cummins, R.A., Walsh, R.N., Budtz-Olsen, O.E., Konstantinas, T., and Horsfall, C.R. 1973. Environmentally induced brain changes in elderly rats. Nature 243, 516-518.

Cummins, R.A., Walsh, R.N., Howard, M.J., and Budtz-Olsen, O.E. 1974b. Hippocampal neuronal development modified by differential rearing. Proc. Aust. Physiol. Pharm. Soc. 20, 30-31.

Cummins, R.A., Worsley, A., Budtz-Olsen, O.E., and Walsh, R.N. 1974a. Testosterone, early experience and behavioral arousal in a novel environment. Horm. Behav. 5, 283-288.

Cummins, R.A., Worsley, A., Budtz-Olsen, O.E., and Walsh, R.N. 1977b. Testosterone and behavior: a reply to Andrew and Archer. Horm. Behav. 8, 123-127.

Cynader, M., Berman, N., and Hein, A. 1976. Recovery of function in cat visual cortex following prolonged deprivation. Exp. Brain Res. 25, 139-156.

Cynader, M., and Charnenko, G. 1976. Abolition of direction selectivity in the visual cortex of the cat. Science 193, 504-505.

Daly, M. 1973. Early stimulation of rodents: A critical review of present interpretations. Brit. J. Psychol. 64, 435.

Daruna, J.H., and Kent, F.W. 1976. Comparison of regional serotonin levels and turnover in the brain of naturally high and low aggressive rats. Brain Res. 101, 489-501.

Darwin, C. 1874. The Descent of Man (2nd ed.) Rand McNally, Chicago, Illinois, 53.

Das, G.D., and Altman, J. 1966. Behavioral manipulations and protein metabolism of the brain. Effects of restricted and enriched environments on the utilization of leucine-$H^3$. Physiol. Behav. 1, 109-110.

Davenport, J.W. 1976. Environmental therapy in hypothyroid and other disadvantaged animal populations, in Environments as Therapy for Brain Dysfunction, Walsh, R., and Greenough, W. eds., Plenum, New York, 71-114.

Davis, W.M., and Brister, C.C. 1971. Increased toxicity of morphine-like analgesics in aggregated mice. J. Pharm. Pharmacol. 23, 882-884.

de Chardin, T. 1965. The Phenomenon of Man. Harper Torchstone, New York.

DeFeudis, F.V. 1971. Effects of environmental changes on the incorporation of carbon atoms of D-glucose into mouse brain and other tissues. Life Sci. 10, 1187-1194.

DeFeudis, F.V. 1972a. Effects of isolation and aggregation on the water content of mouse brain. Biol. Psychiat. 5, 207-210.

DeFeudis, F.V. 1972b. Effects of isolation and aggregation on the incorporation of carbon atoms of D-mannose and D-glucose into mouse brain. Biol. Psychiat. 4, 239-242.

DeFeudis, F.V. 1972c. Specificity of the effects of lithium injections on the entry of carbon atoms of glucose into mouse brain *in vivo*. Arch. Int. Pharmacodyn. 197, 141-146.

DeFeudis, F.V. 1972d. Binding of $^3$H-acetylcholine and $^{14}$C-$\gamma$-aminobutyric acid to subcellular fractions of the brains of differentially housed mice. Neuropharmacology 11, 879-888.

DeFeudis, F.V. 1972e. A note on the subcellular distribution of brain protein in differentially housed mice. Experientia 28, 1427-1428.

DeFeudis, F.V. 1972f. Effects of differential housing on the subcellular distributions of $Li^+$ in the brains of mice. Brain Res. 43, 686-689.

DeFeudis, F.V. 1973. Effects of d-amphetamine on the incorporation of carbon atoms of D-mannose into the brains and sera of differentially housed mice: Short-term reversibility of these effects. Biol. Psychiatry 7, 3-10.

DeFeudis, F.V. 1974a. Differences in brain and body weights of mice caused by differential housing. Experientia 30, 521-522.

DeFeudis, F.V. 1974b. Central Cholinergic Systems and Behavior. Academic Press, London, England, 223pp.

DeFeudis, F.V. 1974c. La biologie de la solitude. Recherche 6, 344-356.

DeFeudis, F.V. 1977. New trends in behavioral neurochemistry, in Behavioral Neurochemistry, Delgado, J.M.R., and DeFeudis, F.V. eds. Spectrum, New York, 11-23.

DeFeudis, F.V., and Black, W.C. 1972. Effects of environment on the incorporation of radioactive carbon atoms of glucose into various structures of the brains of mice; Evidence for diurnal rhythmicity. Exp. Neurol. 36, 41-45.

DeFeudis, F.V., and DeFeudis, P.A. 1977. Elements of the Behavioral Code. Academic Press, London, England, 389 pp.

DeFeudis, F.V., DeFeudis, P.A., and Somoza, E. 1976b. Altered analgesic responses to morphine in differentially housed mice. Psychopharmacologia 49, 117-118.

DeFeudis, F.V., Delgado, J.M.R., and Roth, H. 1970. Content and release of amino acids and catecholamines in monkey brain. Nature (London) 223, 74-75.

DeFeudis, F.V., Delgado, J.M.R., and Roth, H. 1970. Content, synthesis, and collectability of amino acids in various regions of the brains of Rhesus monkeys. Brain Res. 18, 15-23.

DeFeudis, F.V., and Elliott, K.A.C. 1967. Delay or inhibition of convulsions by intraperitoneal injections of diverse substances. Can. J. Physiol. Pharmacol. 45, 857-865.

DeFeudis, F.V., Madtes, P., Ojeda, A., and DeFeudis, P.A. 1976a. "Binding" of $\gamma$-aminobutyric acid and glycine to synaptic particles of the brains of differentially housed mice; evidence for morphological changes. Exp. Neurol. 52, 285-294.

DeFeudis, F.V., and Marks, J.H. 1973a. Studies on the time course of entry and subcellular distribution of radioactivity of $^3H$-d-amphetamine in the brains of differentially housed mice. Experientia 29, 1518-1520.

DeFeudis, F.V., and Marks, J.H. 1973b. Brain to serum distribution of radioactivity of injected $^3H$-d-amphetamine in differentially housed mice. Biol. Psychiat. 85-88.

DeFeudis, F.V., and Paolino, R.M. 1972. Effects of environmental changes on the incorporation of carbon atoms of pyruvate into mouse brain and liver. Experientia 28, 309-310.

DeFeudis, F.V., Paolino, R.M., and DeFeudis, P.A. 1972. Effects of injections of d-amphetamine and lithium chloride on the incorporation of carbon atoms of D-glucose and pyruvate into the brains of isolated and aggregated mice. Fifth Int. Congr. Pharmacol., San Francisco, Calif. p. 53.

DeFeudis, F.V., Somoza, E., DeFeudis, P.A., Pugnaire, M.P., Munoz, L.M., Gonzalez, C., Ibanez, A.E., and Bonnet, K.A. 1978. Environmental model for drug addiction: Studies on the binding of dihydromorphine to synaptic particles of the brains of differentially housed mice, in International Symposium on Factors Affecting the Action of Narcotics, Mario Negri Monograph Series, Adler, M, Manara, L., and Samanin, R. eds., Raven Press, New York, 613-630.

Dejerine, J. 1914. Semiologie des Affections de Systeme Nerveux. Victor Masson et Fils, Paris, France.

del Pozo, F., DeFeudis, F.V., and Jimenez, J. 1978. Environmentally sensitive ultradian motor rhythms in mice. Naturwissenschaften 65, 393.

Denenberg, V.H. 1969. Different effect of pre- and post-weaning stimulation in rats, in Stimulation in Early Infancy, Ambrose, A. ed. Academic Press, New York.

Denenberg, V.H., Woodcoch, J., and Rosenberg, K. 1968. Long-term effects of preweaning and postweaning free-environment experience on rats problem solving behavior. J. Comp. Physiol. Psychol. 66, 533-535.

De Robertis, E. 1971. Molecular biology of synaptic receptors. Science 171, 963-971.

Dewar, A.J., and Winterburn, A.K. 1973a. Changes in the incorporation of C-orotic acid into brain RNA of visually deprived rats following exposure to light. J. Neurol. Sci. 20, 279-285.

Dewar, A.J., and Winterburn, A.K. 1973b. Metabolism of nuclear and cytoplasmic proteins in the visual cortex of sighted rats and rats with retinal degeneration. Exp. Neurol. 41, 584-598.

Dews, P.B., and Wiesel, T.N. 1970. Consequences of monocular deprivation on visual behavior in kittens. J. Physiol. (London) 206, 437-455.

Diamond, M.C. 1967. Extensive cortical depth measures and neuron size increases in the cortex of environmentally enriched rats. J. Comp. Neurol. 131, 357-364.

Diamond, M.C. 1976. Anatomical brain changes induced by environment, in Knowing, Thinking and Believing, Petrinovich, L. and McGaugh, J. eds., Plenum Press, New York, 215-239.

Diamond, M.C., Ingham, C., Johnson, R., Bennett, E., and Rosenzweig, M. 1976. Effects of environment on morphology of rat cerebral cortex and hippocampus. J. Neurobiol. 7, 75-85.

Diamond, M.C., Johnson, R., and Ingham, C.A. 1971. Brain plasticity induced by environment and pregnancy. Internat. J. Neurosci. 2, 171-178.

Diamond, M.C., Johnson, R., Ingham, C., Rosenzweig, M.R., and Bennett, C. 1975. Effects of differential experience on neuronal nuclear and perikarya dimensions in the rat cerebral cortex. Behav. Biol. 15, 107-111.

Diamond, M.C., Krech, D., and Rosenzweig, M.R. 1964. The effects of an enriched environment on the histology of the rat cortex. J. Comp. Neurol. 123, 111-120.

Diamond, M.C., Lindner, B., Johnson, R., Bennett, E., and Rosenzweig, M. 1975. Differences in occipital cortex synapses from enriched, imp, over, and standard colony rats. J. Neurosci. Res. 1, 109-119.

Diamond, M.C., Rozenzweig, M.R., Bennett, E.L., Lindner, B., and Lyon, L. 1972. Effects of environmental enrichment and impoverishment on rat cerebral cortex. J. Neurobiol. 3, 47-64.

Diamond, M.C., Rosenzweig, M.R., and Krech, D. 1965. Relationships between body weight and skull development in rats raised in enriched and impoverished conditions. J. Exp. Zool. 160, 29-36.

Diamond, M.C., Law, F., Rhodes, H., Lindner, B., Rosenzweig, M.R., Krech, D., and Bennett, E.L. 1966. Increases in cortical depth and glial numbers in rats subjected to enriched environment. J. Comp. Neurol. 128, 117-126.

Diaz, J., Ellison, G., and Masuoka, D. 1975. CNS response to injury in enriched and impoverished environments: A behavioral and neurochemical study in plasticity. Soc. Neurosci. Abst. 1, 511.

Diaz, J., Ellison, G., and Masouka, D. 1978. Stages of recovery from central norepinephrine lesions in enriched and impoverished environments: A behavioral and biochemical study. Exp. Brain Res. 31, 117-130.

Dismukes, R.K., and Daly, J.W. 1976. Altered brain cycle, AMP-responses in rats reared in enriched or impoverished environments. Experientia 32, 730-732.

Doemin, N.N. 1968. Effects of acetylcholine and adrenaline upon some sites of nucleic acid metabolism in single cells and subcellular fractions of nervous tissue, in Macromolecules and the Function of the Neuron, Lodin. Z., and Rose, S.P.R. eds., Excerpta Medica Foundation, Amsterdam, Holland, 316-323.

Donaldson, H.H. 1890. Anatomical observations on the brain and several sense organs of the blind deaf mute, Laura Derney Bridgeman, I. Am. J. Psychol. 3, 293-342.

Donaldson, H.H. 1891. Anatomical observations on the brain and several sense organs of the blind deaf mute, Laura Derney Bridgeman, II. Am. J. Psychol. 4, 248-294.

Dorfman, A., and Goldbaum, L.R. 1947. Detoxification of barbiturates J. Pharmocol. Exp. Ther. 90, 330-337.

Dru, D., Walker, J.P., and Walker, J.B. 1975. Self produced locomotion restores visual capacity after striate lesions. Science 187, 265-266.

Duffy, F. 1957. The psychological significance of the concept of "arousal" or activation. Psychol. Rev. 64, 265-275.

Earley, C.J., and Leonard, B.E. 1977. The effect of testosterone and cyproterone acetate on the concentration of $\gamma$-aminobutyric acid in brain areas of aggressive and non-aggressive mice. Pharmacol. Biochem. Behav. 6, 409-413.

Eayrs, J.T., and Goodhead, B. 1959. Postnatal development of the cerebral cortex in the rat. J. Anat. 93, 385-402.

Edwards, H., Barry, W., and Wyspianski, J. 1969. Effect of differential rearing on photic evoked potentials and brightness discrimination in the albino rat. Dev. Psychobiol. 2, 133-138.

Einon, D., Stewart, J., Atkinson, S., and Morgan, M. 1976. Effect of isolation on barbiturate anaesthesia in the rat. Psychopharmacologia 50, 85-88.

Einstein, A. 1923. The Principle of Relativity. Dover, New York.

Eliot, C. 1969. Japanese Buddhism. Barnes and Nobel, New York.

Ellison, G., and Byran, K. 1976. Evidence that supersensitivity after recovery from 6-hydroxydopamine is an artifact of isolation housing. Neurosci. Letters 3, 229-232.

Entingh, D., Dunn, A., Glassman, E., Wilson, J., Hogan, E., and Damstra, T. 1975., in Handbook of Psychobiology, Gassaniga, M. and Blakemore, C., eds. Academic Press, New York, 201-235.

Erhard, W. , Gioscia, V., and Anbender, K. 1978. Being well. in Beyond Health and Normality: Explorations of Extreme Psychological Wellbeing. Van Nostrand Reinhold, New York (In press).

Essman, W.B. 1968. Differences in locomotor activity and brain serotonin metabolism in differentially housed mice. J. Comp. Physiol. Psychol. 66, 244-246.

Essman, W.B. 1969. Free and motivated behavior and amine metabolism in isolated mice, in Aggressive Behavior. Garattini, S., and Sigg, E.B. eds., John Wiley and Sons, New York 203-208.

Essman, W.B. 1971a. Neurochemical changes associated with isolation and environmental stimulation. Biol. Psychiat. 3, 141-147.

Essman, W.B. 1971b. Isolation-induced behavioral modification: Some neurochemical correlates, in Brain Development and Behavior. Academic Press, New York 265-276.

Essman, W.B. 1972. Contributions of differential housing to brain development: Some implications for sleep behavior, in Sleep and the Maturing Nervous System, Clemente, D.D., Purpura, D.P., and Mayer, F.E. eds., Academic Press, New York 99-107.

Essman, W.B., and Smith, G.E. 1967. Behavioral and neurochemical differences between differentially housed mice. Amer. Zool. 7, 792.

Eterovic, V.A., and Ferchmin, P.A. 1974. Interaction of environment and injections on brain weight in rats. Dev. Psychobiol. 7, 515-517.

Faure, C., and Soula, C. 1913. Comp. Rend. Soc. Biol. 74, 350.

Feinberg, I. 1969. Effects of age on human sleep patterns, in Sleep: Physiology and Pathology, Kales, A. ed. Lippincott, Philadelphia 39-52.

Feinberg, I., Brain, M., and Shulman, E. 1969. EEG sleep paterns in mental retardation. Electroencephalogr. Clin. Neurophysiol. 27, 128-141.

Ferchmin, P.A., Bennett, E.L., and Rosenzweig, M.R. 1975. Direct contact with enriched environments is required to alter cerebral weights in rats. J. Comp. Physiol. Psychol. 88, 360-367.

Ferchmin, P.A., Eterovic, V., and Caputto, R. 1970. Studies of brain weight and RNA content after short periods of exposure to environmental complexity. Brain Res. 20, 49-57.

Ferchmin, P., and Eterovic, V.A. 1973. Personal communication.

Ferchmin, P., and Eterovic, V.A. 1977. Brain plasticity and environmental complexity: Role of motor skills. Physiol. Behav. 18, 455-461.

Ferguson, J., and Dement, W. 1969. The behavioral effects of amphetamine on REM deprived rats. J. Psychiat. Res. 7, 111-118.

Fertig, A., Kierman, J.A., and Seyan, S. 1971. Enhancement of axonal regeneration in the brain of the rat by corticotrophin. Exp. Neurol. 33, 372-385.

Fiala, B.A., Joyce, J., and Greenough, W.T. 1978. Environmental complexity modulates growth of granule cell dendrites in developing but not adult hippocampus of rats. Exp. Neurol. 59, 372-383.

Fiala, B.A., Snow, F., and Greenough, W.T. 1977. "Impoverished" rats weigh more than "enriched" rats because they eat more. Dev. Psychobiol. 10, 537-541.

Fifkova, E. 1968. Changes in the visual cortex of rats after unilateral deprivation. Nature 220, 379-381.

Fifkova, E. 1970a. The effect of monocular deprivation on the synaptic contacts of the visual cortex. J. Neurobiol. 1, 285-294.

Fifkova, E. 1970b. Changes in axosomatic synapses in the visual cortex of monocularly deprived rats. J. Neurobiol. 2, 61-71.

Fifkova, E., and Hassler, R. 1969. Quantitative morphological changes in visual centres in rats after unilateral deprivation. J. Comp. Neurol. 135, 167-178.

Fog, R., Randrup, A., and Pakkenberg, H. 1970. Lesions in corpus striatum and cortex of rat brains and the effect on pharmacologically induced stereotyped, aggressive and cataleptic behavior. Psychopharmacologia 18, 346-356.

Fore, M.W. 1967. The effects of short term social and sensory isolation upon behavior, EEG and averaged evoked potentials in puppies. Physiol. Behav. 2, 145-151.

Forgays, D.G., and Forgays, J.W. 1952. The nature of the effect of free-environmental experience in the rat. J. Comp. Physiol. Psychol. 45, 322-328.

Forgus, R.H. 1954. The effect of early perceptual learning on the behavior organization of adult rats. Physiol. Psychol. 47, 331-336.

Forster, M., and Sherrington, C.S. 1897. A Textbook of Physiology. Part III. The Central Nervous System. MacMillan, New York, 1117.

Fourment, A., and Scherrer, J. 1961. Résponses electrocorticales du lapin élevé dans l'obscurité. J. Physiol. (Paris) 53, 340-351.

Fox, M.W. 1967. The effects of short term social and sensory isolation upon behavior, EEG, and average evoked potentials in puppies. Physiol. Behav. 2, 145-151.

Freeman, D. 1971. Aggression: Instinct or symptom? Aust. N.Z. J. Psychiat. 5, 66-73.

Friedman, A.H. and Walker, C.A. 1968. Circadian rhythms: Rat midbrain and caudate nucleus biogenic amine levels. J. Physiol. (London) 197, 77-86.

Fuller, J.L. 1967. Experiential deprivation and later behavior. Science 158, 1645-1652.

Fuller, R.W., and Hines, C.W. 1967. d-Amphetamine levels in brain and other tissues of isolated and aggregated mice. Biochem. Pharmacol. 16, 11-16.

Ganz, L. 1975. Orientation in visual space by neonates and its modification by visual deprivation, in The Developmental Neuropsychology of Sensory Deprivation, Riesen, A.H. ed., Academic Press, New York.

Ganz, L., Fitch, M., and Satterberg, G.A. 1968. Selective effect of visual deprivation on receptive field shape determined neurophysiologically. Exp. Neurol. 22, 614-637.

Ganz, L., and Haffner, M.F. 1974. Permanent perceptual and neurophysiological effects of visual deprivation in the cat. Exp. Brain Res. 20, 67-87.

Garattini, S., Giacalone, E., and Valzelli, L. 1967. Isolation, agressiveness and brain 5-hydroxytryptamine turnover. J. Pharm. Pharmacol. 19, 338-339.

Garattini, S., Giacalone, E., and Valzelli, L. 1969. Biochemical changes during isolation-induced aggressiveness in mice, in Aggressive Behaviour, Garattini, S., and Sigg, E.B. eds., Excerpta Medica Fdn., Amsterdam, Holland, 179-187.

Garattini, S., and Valzelli, L. 1965. Serotonin. Elsevier Publ. Co. Amsterdam, Holland.

Gardner, F., Debold, R., Fershein, W., and Heermans, H. 1970. Increased incorporation of $^{14}C$-uridine into rat brain RNA as a result of novel electroshock. Nature (London) 227, 1242-1243.

Garey, L.J., and Dursteler, M.R. 1975. Reversal of deprivation effects in the lateral geniculate of the cat. Neurosci. Letters 1, 19-24.

Gazzaniga, M.S. 1974. Determinants of cerebral recovery, in Plasticity and Recovery of Function in the Central Nervous System, Stein, D.G., Rosen, J.T., and Butters, N. eds., Academic Press, New York.

Geinismann, Y.Y. 1971. Nucleic acid content of spinal motorneurons and their satellites under orthodromic and antidromic stimulation. A cytospectrophotometric study. Brain Res. 28, 251-262.

Geinismann, Y.Y., Larina, V.N., and Mats, V.N. 1971. Changes of neurones' dimensions as a possible morphological correlate of their increased functional activity. Brain Res. 26, 247-257.

Geller, E. 1971. Some observations on the effect of environmental complexity and isolation in biochemical ontogeny, in Brain Development and Behavior, Sluman, M.B., McGinty, D.H., and Adenolfi, A.M. eds., Academic Press, New York, 277-296.

Geller, E., and Yuwiler, A. 1968. In Ontogenesis of the Brain, Jilek, L., and Trojan, S. eds., Charles University Press, Prague, 277-284.

Geller, E., Yuwiler, A., and Zolman, J. 1965. Effects of environmental complexity on constituents of brain and liver. J. Neurochem. 12, 949-955.

Gendreau, P., Freedman, N., Wilde, G., and Scott, G. 1972. Changes in EEG alpha frequency and evoked response latency during solitary confinment. J. Abnorm. Psychol. 79, 54-59.

Geschwind, N. 1974. Late changes in nervous system: An overview. In, Plasticity and Recovery of Function in the Central Nervous System, Stein, D.G., Rosen, J.J., and Butters, N. eds., Academic Press, New York, 467-508.

Giacobini, E. 1959. The distribution and localization of cholinesterase in nerve cells. Acta Physiol. Scand. 45, (suppl. 156), 80.
Giacalone, E., Tansella, M., Valzelli, L., and Garattini, S. 1968. Brain serotonin metabolism in isolated aggressive mice. Biochem. Pharmacol. 17, 1315-1327.
Gisiger, V. 1971. Triggering of RNA synthesis by acetylcholine stimulation of the postsynaptic membrane in a mammalian sympathetic ganglion. Brain Res. 33, 139-146.
Glassman, E. 1969. RNA and learning. Ann. Rev. Biochem. 38, 605-646.
Glassman, E., Gesper, W., Perumal, R., Malchus, B., and Wilson, J. 1973. The effect of short experiences on the incorporation of radioactive phosphate into synaptosomal and non-histone acid-extractable nuclear proteins from rats and mouse brain. Int. Cong. Pharmacol., Fifth 4, 14-17.
Glassman, E., and Wilson, J.E. 1970. The incorporation of uridine into RNA during short experiences. Brain Res. 21, 157-168.
Glees, P., and Le Gros Clark, W. 1941. The termination of optic fibres in the lateral geniculate body of the monkey. J. Anat. 75, 295-308.
Globus, A., Rosenzweig, M.R., Bennett, E., and Diamond, M. 1973. Effects of differential experience on dendritic spine counts in rat cerebral cortex. J. Comp. Physiol. Psychol. 82, 175-181.
Globus, A., and Scheibel, A. 1966. Loss of dendritic spines as an index of presynaptic terminal patterns: An experimental application of the golgi method. Nature (London) 212, 463-465.
Globus, A., and Scheibel, A. 1967a. The effect of visual deprivation on cortical neurons: A golgi study. Exp. Neurol. 19, 331-345.
Globus, A., and Scheibel, A. 1967b. Synaptic loci in cortical neuron of the rabbit: The specific afferent radiation. Exp. Neurol. 18, 116-131.
Globus, G., and Franklin, S. 1976. "Thought and meditation." Unpublished manuscript.
Goldberg, A.M., and Welch, B.L. 1972. Adaptation of the adrenal medulla: Sustained increase in choline acetyltransferase by psychosocial stimulation, Science 178, 319-320.
Goldberger, M.E. 1974. Recovery of movement after CNS lesions in monkeys, in Plasticity and Recovery of Function in the Central Nervous System, Stein, D.G., Rosen, H.J., and Butters, N. eds., Academic Press, New York, 265-338.
Goldman, P.S. 1976. The role of experience in recovery of function following orbital prefrontal lesion in infant monkeys. Neuropsychologia 14, 401-412.
Goldman, P.S., and Mendelson, M.J. 1977. Salutary effects of early experience on deficits caused by lesions on frontal associational cortex in developing rhesus monkeys. Exp. Neurol. 57, 588-602.
Goldstein, J. 1976. The Experience Of Insight. Unity Press, Santa Cruz, California.
Goleman, D. 1977. The Varieties of The Meditative Experience. Dutton, New York.
Goleman, D. 1978. A map of inner space. in Beyond Health and Normality: Explorations of Extreme Psychological Wellbeing, Walsh, R., and Shapiro, D. eds., Van Nostrand Reinhold, New York, (In press).
Gombos, G., Morgan, I.G., Waehneldt, T.V., Vincendon, G., and Breckenridge, W.C. 1972. Glycoproteins of the synaptosomal plasma membrane. Adv. Exp. Med. Biol. 25, 101-113.
Gomirato, G. 1954. J. Neuropathol. Exp. Neurol. 21, 634.
Goodman, L. 1932. Effects of total absence of function on the optic system of rabbits. Am. J. Physiol. 100, 46-63.
Gottschalk, L., Hoer, J., and Bates, D. 1973. Effects of sensory overload on psychological state. Arch. Gen. Psychiatry 27, 451.458.

Govinda, L.A. 1969. Foundations of Tibetan Mysticism. Samuel Weiser, New York.

Grafstein, B., and Sastry, P. 1957. Some preliminary physiological studies on chronic neuronally isolated cerebral cortex. Electroencephalogr. Clin. Neurophysiol. 9, 723-725.

Gray, E.G. 1959. Axosomatic and axodendritic synapses of the cerebral cortex: an electron microscopic study. J. Anat. 93, 420-432.

Greenough, W.T. 1976. Enduring brain effects of differential experience and training, in Neural Mechanisms of Memory and Learning, Rosenzweig, M., and Bennett, E. eds., MIT Press, Cambridge, Mass. 255-278.

Greenough, W.T., Fass, B., and DeVoogd, T.J., 1976. The influence of experience on recovery following brain damage in rodents: Hypotheses based on developmental research, in Environments as Therapy for Brain Dysfunction, Walsh, R., and Greenough, W.T. eds., Plenum Press, New York, 10-50.

Greenough, W.T., and Maier, S.F. 1972. Molecular changes during learning: Behavioral strategy - a comment on Gaito and Bonnett. Psychol. Bull. 78, 480-482.

Greenough, W.T., and Teyler, T. 1978. Development and memory: the synaptic connection, in Brain and Learning, Teyler, T. ed., Greylock Publishers, Stamford, Connecticut.

Greenough, W.T., Volkmar, R., and Juraska, M. 1973. Effects of rearing complexity on dendritic branching in frontolateral and temporal cortex of the rat. Exp. Neurol. 41, 371-378.

Greenough, W.T., West, R.W., and DeVoogd, T.J. 1978. Subsynaptic plate perforations: Changes with age and experience in the rat. Science 202, 1096-1098.

Griffit, W., and Veitch, R. 1971. Hot and crowded: Influences of population density and temperature on interpersonal affective behavior. J. Personal. Soc. Psychol. 17, 92-98.

Grouse, L., Schrier, B.K., Bennett, E., Rosenzweig, M., and Nelson, P. 1978. Effects of environmental complexity on rat brain RNA diversity. J. Neurochem. 30, 191-203.

Guenther, H. 1976. Philosophy and Psychology in the Abhidharma. Shambala, Berkeley, California.

Guenther, H., and Kawamura, L.S., trans. 1975. Mind in Buddhist Psychology. Dharma Press, Emeryville, California.

Guillery, R.W. 1972. Binocular competion in the control of geniculate cell growth. J. Comp. Neurol. 144, 117-130.

Gunn, J.A., and Gurd, M.R. 1940. The action of some amines related to adrenaline, cyclohexylalkylamines. J. Physiol. (London) 97, 453-470.

Guth, L. 1974. Axonal regeneration and functional plasticity in the central nervous system. Exp. Neurol. 45, 606-654.

Gyllensten, L., Malmfors, T., and Norrlin, M. 1965. Effect of visual deprivation on the optic centres of growing and adult mice. J. Comp. Neurol. 124, 149-160.

Gyllensten, L., Malmfors, T., and Norrlin, M. 1966. Growth alteration in the auditory cortex of visually deprived mice. J. Comp. Neurol. 126, 463-470.

Haggert, B.N., and Werboff, J. 1968. Temperature reduction in neonatal mice: Effects on later behavior. Dev. Psychobiol. 1, 257-266.

Hamlin, R.N. 1970. Intellectual functions 14 years after frontal lobe surgery. Cortex 6, 299-307.

Harlow, H.F., and Harlow, M.K. 1962. The effects of rearing conditions on behavior. Bull. Menninger Clin. 26, 213-224.

Harrison, R. 1904. An experimental study of the relation of the nervous system to the developing musculature in the embryo of the frog. Amer. J. Anat. 3, 197-220.

Hatch, A., Balazs, T., Wieberg, G., and Grice, S.H. 1963. Long-term isolation stress on mice. Science 142, 507.

Hawley, A. 1950. Human Ecology. Ronald Press, New York, 3.

Haywood, J., Rose, S., and Bateson, P. 1970. Effects of an imprinting procedure on RNA polymerase activity in the chick brain. Nature (London) 228, 373-374.

Hebb, D.O. 1947. The effects of early experience on problem solving at maturity. Am. Psychol. 2, 306-307.

Hebb, D.O. 1949. The Organization of Behavior. Wiley, New York.

Hebb, D.O. 1955. Drives and the C.N.S. (Conceptual Nervous System). Psychol. Rev. 62, 243-254.

Hebb, D.O., and Thompson, W.R. 1954. The social significance of animal studies, in Handbook of Social Psychology. Lindzey, G. ed., Addison-Wesley, Reading, Mass.

Hein, A. 1970. Recovering spatial motor coordination after visual cortex lesion, in Perception and Its Disorders, Hamburg, D.A., Pribram, K.H. eds., Williams and Wilkins, Baltimore, 163-175.

Heisenberg, W. 1958. Physics and Philosophy. Harper and Row, New York.

Heisenberg, W. 1963. Physics and Philosophy. Allen and Unwin, London, England.

Henderson, N.D. 1969. Prior treatment effects on open field emotionality: The need for representative design, in Experimental Approaches to the Study of Emotional Behavior, Tobach, E. ed. Annals N.Y. Acad. Sci. 159, 860-868.

Henderson, N. 1970a. Brain weight increases resulting from environmental enrichment: a directional dominance in mice. Science 169, 776-778.

Henderson, N.D. 1970b. Genetic influences on the behavior of mice can be obscured by laboratory rearing. J. Comp. Physiol. Psychol. 72, 505-511.

Henderson, N.D. 1973. Brain and body weight changes resulting from enriched rearing conditions: A diallel analysis. Dev. Psychobiol. 6, 367-376.

Hendley, E.D., Moisset, B., and Welch, B.L. 1973. Catecholamine uptake in cerebral cortex: Adaptive change induced by fighting. Science 180, 1050-1052.

Henneman, D.H., Altschule, M.D., and Gonca, R.M. 1954. Carbohydrate metabolism in brain disease. II. Glucose metabolism in schizophrenic, manic-depressive and involutional psychoses. Arch. Intern. Med. 94, 402-416.

Hernandez-Peon, R., Chavez-Ibarra, G, Morgane, P., and Timo-Iaria, C. 1963. Limbic cholinergic pathways involved in sleep and emotional behavior. Exp.Neurol. 8, 93-111.

Hess, J.L., Denenburg, V., Zarrow, M., and Pfiefer, W. 1969. Modification of the corticosterane response as a function of handling in infancy. Physiol. Behav. 4, 109-111.

Hicks, S.P., and D'Amato, C.J. 1970. Motor-sensory behavior after hemispherectomy in newborn and mature mice. Exp. Neurol. 29, 416-438.

Hilgard, E.R., and Bower, G.H. 1975. Theories of Learning (4th ed.). Prentice-Hall, Englewood Cliffs, New Jersey.

Hillarp, N.A., Fuxe, K., and Dahlstom, A. 1966. Demonstration and mapping of central neurones containing dopamine, noradrenaline, and 5-hydroxytryptamine and their reactions to psychopharmaca. Pharmacol. Rev. 18, 727-742.

Himwich, W.A. 1969. The effect of environment upon the developing brain, in The Future of the Brain Sciences. Plenum Press, New York, 237-252.

Hobbes, T. 1650. Human Nature. Reproduced in The English Works of Thomas Hobbes (4 Vols.), Molesworth, W. ed., Bohn, London (1839).

Holdstock, T.L., and Franks, P.E. 1971. Some aspects of paradoxical sleep in the laboratory rat. Psychologica Africana 14, 38-44.

Holloway, R.L. 1966. Dendritic branching: some preliminary results of training and complexity in rat visual cortex. Brain Res. 2, 393-396.

Hoover, D., and Diamond, M.C. 1976. The effects of norethynodrel administration on the rat visual cortex exposed to differential environment: A preliminary study of electrolytes and water. Brain Res. 103, 139-142.

Horn, G. 1965. The effect of somesthetic and photic stimuli on the activity of units in the striate cortex of unanaesthetized unrestrained cats. J. Physiol. 179, 263-277.

Hoyle, F. 1960. The Nature of the Universe. Harper and Row, New York.

Hubel, D.H., and Wiesel, T.N. 1962. Receptive fields, binocular interaction and functional architecture in the cats' visual cortex. J. Physiol. 160, 106-154.

Hubel, D.H., and Wiesel, T.N. 1970. The period of susceptibility to the physiological effects of unilateral eye closure in kittens. J. Physiol. 206, 419-436.

Hughes, A. 1968. Development of limb innervation, in Growth of the Nervous System, Westernholme, G.E.W., and O'Connor, M. eds., A Ciba Foundation Symposium. Little, Brown and Co., Boston, Mass.

Hughes, J. 1975. Isolation of an endogenous compound from the brain with pharmacological properties similar to morphine. Brain Res. 88, 295-308.

Hunt, J. McV. 1976. Environmental programming to foster competence and prevent mental retardation in infancy, in Environments as Therapy for Brain Dysfunction, Walsh, R:N., and Greenough, W.T. eds. Plenum Press, New York, 201-255.

Huntley, M.J., and Newton, J.M. 1972. Effects of environmental complexity and locomotor activity on brain weight in the rat. Physiol. Behav. 8, 725-727.

Hutchings, C. 1967. Infantile body temperature loss in rats: Effects of duration of cold exposure, rate of heat loss, level of hypothermia, and rewarming on later emotionality. Psychol. Rep. 21, 985-1002.

Hyden, H. 1943. Protein metabolism in the nerve cell during growth and function. Physiol. Scand. 6 (suppl. 17), 3-136.

Hyden, H. 1967. Biochemical and molecular aspects of learning and memory. Proc. Am. Phil. Soc. 111, 326-342.

Hyden, H., and Lange, P.W. 1965. Proc. Nat. Acad. Sci. USA 53, 946.

Hyden, H., and Lange, P.W. 1972. Protein changes in different brain areas as a function of intermittent training. Proc. Nat. Acad. Sci. (Washington, D.C.) 69, 1980-1984.

Hymovich, B. 1952. The effects of experimental variations on problem solving in the rat. J. Comp. Physiol. Psychol. 45, 313-321.

Isaac, W. 1964. Role of stimulation and time in the effects of spaced occipital ablations. Psychol. Rep. 14, 151-154.

Isaacson, R.L. 1974. Recovery "?" from early brain damage. Paper presented at the National Conference on Early Intervention with High Risk Infants and Young Children. Chapel Hill, North Carolina.

Ivanow-Smolenski, A.G. 1954. Srundzuge der pathosphsiologie er holeren Nerventatigkeit. Akademic Verlag. Berlin, Germany.

Jacobson, M. 1970. Developmental Neurobiology. Holt, Rinehart and Winston, New York.

Jacobson, S. 1963. Sequence of myelinization in the brain of the albino rat. A: Cerebral cortex, thalamus and related structures. J. Comp. Neurol. 121, 5-29.

Janssen, P.A.J., Jagenau, A.H., and Niemegeers, C.S.E. 1960. Effects of various drugs on isolation-induced fighting behavior in male mice. J. Pharmacol. Exp. Ther. 129, 471-475.

Jessor, R., and Richardson, S. 1968. Psychosocial deprivation and personality development: Overview. In, Perspectives on Human Deprivation: Biological Psychological and Sociological. U.S. Department of Health, Education and Welfare, Washington, D.C.

Jones, S.J., and Moss, H.A. 1971. Age, state and maternal behavior associated with infant vocalization. Child Dev. 42, 1039-1051.

Jung, R. Kornhuber, H.H., and laFonesca, J.S. 1963. Multisensory convergence on cortical neurons. Neuronal effects of visual, acoustic and somesthetic stimuli in the superior convolutions of the In Progress in Brain Research 1, Brain Mechanisms, Morrizzi, G., Fessard, A., and Jasper, H.H. eds., Amsterdam, Elsevier 207-234.

Kalil, R.E. 1973. Formation of new retino-geniculate connections in kittens: Effects of age and visual experience. Anat. Rec. 175, 353.

Kant, I. 1781. Kritik der remein Vermunft. P. Reclam, Leipzig. English ed: Critique of Pure Reason, 1887. J. Meiklejohn, trans. George Bell, London, England.

Kapleau, P. 1967. The Three Pillars of Zen. Beacon Press, Boston, Mass.

Karlu, Rinpoche. 1977. Cited by J. Goldstein. Talk given at Insight Meditation Center, Barre, Massachusetts, November.

Kasamatsu, T. 1976. A long lasting change in ocular dominance of kitten striate neurons induced by reversible unilateral blockade of tonic retinal discharges. Exp. Brain Res. 26, 487-494.

Katz, D.M., and Steinberg, H. 1972. Factors which might modify morphine dependence in rats, in, Biochemical and Pharmacological Aspects of Dependence and Reports on Marihuana Research, Van Praag, H.N, De Erven, F., and Bohn, N.V. eds. pp. 46-61.

Kernell, D., and Peterson, R. 1970. The effect of spike activity versus synaptic activation on the metabolism of ribonucleic acid in a molluscan giant neurone. J. Neurochem. 17, 1087-1094.

Kety, S.S., Javoy, F., Thierry, A.M., Julou, L., and Glowinski, J. 1967. A sustained effect of electro-convulsive shock on the turnover of norepinephrine in the central nervous system of the rat. Proc. Nat. Acad. Sci. USA 58, 1249-1254.

Kimble, D.P. 1963. The effects of bilateral hippocampal lesions in rats. J. Comp. Physiol. Psychol. 56, 273-283.

Kleitman, N. 1963. Sleep and Wakefulness. University of Chicago Press, Chicago, Illinois.

Koelle, G.B. 1952. Histochemical localization of cholinesterases in the central nervous system of the rat. J. Pharmacol. Exp. Ther. 106, 401.

Koelle, G.B. 1954. The histochemical localization of cholinesterase in the central nervous system of the rat. J. Comp. Neurol. 100, 211.

Konrad, K., and Melzack, R. 1975. Novelty enhancement effects associated with early sensory-social isolation, in The Developmental Neuropsychology of Sensory Deprivation, Riesen, A.H. ed., Academic Press, New York, 253-276.

Kooy, F.H. 1919. Hyperglycemia in mental disorders. Brain 42, 214-289

Korner, A ., Kraemer, H., Jaffner, M., and Cooper, L. 1975. Effects of waterbed flotation on premature infants: A pilot study. Pediatrics, 56.

Korner, A., and Thoman, E.B. 1972. The relative efficacy of contact and vestibular proprioceptive stimulation in soothing neonates. Child Dev. 43, 443-453.

Kosower, E.M. 1972. A molecular basis for learning and memory. Proc. Nat. Acad. Sci. (Washington, D.C.) 69, 3292-3296.

Kostowski, W., Czlonkowski, A., Rewerski, W., and Piechocki, T. 1977. Morphine action in grouped and isolated rats and mice. Psychopharmacologia 53, 191-193.

Kostowski, W., Giacolone, E., Garattini, S., and Valzelli, L. 1968. Studies on behavioral and biological changes in rats after lesion of midbrain raphe. Eur. J. Pharmacol. 4, 371-376.

Kratz, K.E., Spear, P.D., and Smith, D.C. 1976. Post-critical period reversal of effects of monocular deprivation on striate cortex cells in the cat. J. Neurophysiol. 39, 501-511.

Krech, D., Rosenzweig, M., and Bennett, E.L. 1960. Effects of environmental complexity and training on brain chemistry. J. Comp. Physiol. Psychol. 53, 509-519.

Krech, D., Rosenzweig, M.R., and Bennett, E.L. 1962a. Relations between brain chemistry and problem solving among rats in enriched and impoverished environments. J. Comp. Physiol. Psychol. 55, 801-807.

Krech, D., Rosenzweig, M.R., and Bennett, E.L. 1962b. Effects of environmental complexity and training on brain chemistry. J. Comp. Physiol. Psych. 53, 509-519.

Krech, D., Rosenzweig, M., and Bennett, E.L. 1963. Effects of complex environment and blindness on rat brain. Arch. Neurol. 8, 403-412.

Krech, D., Rosenzweig, M.R., and Bennett, E.L. 1966. Environmental impoverishment, social isolation, and changes in brain chemistry and anatomy. Physiol. Behav. 1, 99-106.

Kreig, W.J.S. 1946. Connections of the cerebral cortex, J. Comp. Neurol. 84-85, 277-323.

Kuenzle, C.C. and Knüsel, A. 1974. Mass training of rats in a superenriched environment. Physiol. Behav. 13, 205-210.

Kuhlenkampf, H. 1952. Das Verhalten der Neuroglia in den Vordenhornan des Ruckenmarkes der weissen maus unter dem Reiz physiologischer Tatigkeit. J. Anat. Entw. 116, 304-312.

Kuhn, T.S. 1970. The Structure of Scientific Revolutions (2nd ed.). University of Chicago Press, Chicago, Illinois.

Lacey, J.I. 1967. Somatic response patterning and stress: Some revisions of activation theory, in Psychological Stress, Appley, M.H. and Trumball, R., eds. Appleton-Century-Crofts, New York.

Lal, H., DeFeo, J.J., Pitterman, A., Patel, G., and Baumel, I. 1972. Effects of prolonged social deprivation or enrichment on neuronal sensitivity for CNS depressants and stimulants, in Drug Addiction: Experimental Pharmacology, Vol. 1. Futura Publishing Co., Mount Kisco, New York, 255-266.

Lambert, J.F., and Truong-Ngoc, A. 1976a. Influence de l'environment instrumental et social sur l'excitabilité nerveuse centrale de rat Wistar male: Misure de la reactivité du systeme reticulo-cortical. Aggressologie 17, 13-18.

Lambert, J.F., and Truong-Ngoc, A. 1976b. Influence de l'environment instrumental et social sur la structure d'un echantillon du cycle veille-sommeil chez le rat Wistar male: Correlations avec les modifications de l'excitabilite du systeme reticulo-cortical. Aggressologie 17, 19-25.

Lashley, K.S. 1943. Studies of cerebral functions in learning XII -- loss of the maze habit after occipital lesions in blind rats. J. Comp. Neurol. 79, 431-472.

LaTorre, J.C. 1968. Effect of differential environmental enrichment on brain weight and on acetylcholinesterase and cholinesterase activities in mice. Exp. Neurol. 22, 493-503.

Leff, J.P. 1968. Perceptual phenomena and personality in sensory deprivation. Br. J. Psychiat. 114, 1499-1508.

Le Gros Clark, W., and Penman, G. 1933-1934. The projection of the retina in the lateral geniculate body. Proc. R. Soc. (London Ser. B) 114, 291-313.

LeShan, L. 1969. Physicists and mystics: similarities in world view. J. Transp. Psychol. 1(2), 1-20.

Leuba, C. 1955. Toward some integration of learning theories: The concept of optimal stimulation. Psychol. Rep. 1, 27-33.

LeVere, T.E. 1975. Neural stability, sparing and behavioral recovery following brain damage. Psych. Rev. 82, 344-358.

LeVere, T.E., and Davis, N. 1977. Recovery of function after brain damage: The motivational specificity of spared neural traces. Exp. Neurol. 57, 883-900.

Levine, S. 1962. The effects of infantile experience on adult behavior, in Experimental Foundations of Clinical Psychology, Bachrach, A.J., ed., Basic Books, New York, 139-169.

Levine, S. 1969a. Infantile stimulation: A perspective, in Stimulation in Early Infancy, Ambrose, A. ed. Academic Press, London, England.

Levine, S. 1969b. An endocrine theory of infantile stimulation, in Stimulation in Early Infancy, Ambrose, A., ed., Academic Press, New York.

Levine, S., Chevalier, J.A., and Kerchin, S.A. 1956. The effects of shock and handling in infancy on later avoidance learning. J. Personal. 24, 475-493.

Levine, S., and Wiener, S. 1976. Malnutrition and early environmental experience: Possible interactive effects on later behavior, in Environments as Therapy for Brain Dysfunction, Walsh, R., and Greenough, W. eds., Plenum Press, New York, 51-70.

Levitan, I.B., Mushynski, W.E., and Ramirez, G. 1972a. Effects of an enriched environment on amino acid incorporation into rat brain subcellular fractions *in vivo*. Brain Res. 41, 498-502.

Levitan, I.B., Mushynski, W.E., and Ramirez, G. 1972b. Effects of environmental complexity on amino acid incorporation into rat cortex and hippocampus *in vivo* J. Neurochem. 19, 2621-2630.

Levitan, I.B., Ramirez, G., and Mushynski, W.E. 1972c. Amino acid incorporation in the brains of rats trained to use the non-preferred paw in retrieving food. Brain Res. 47, 147-156.

Levy, D.N. 1934. Experiments on the sucking reflex and social behavior in dogs. Am. J. Orthopsychiatry 4, 203-224.

Levy, D. 1938. On instinct satiation: An experiment on the pecking behavior of chickens. J. Gen. Psychol. 18, 227-248.

Lewis, D.J. 1969. Selective amnesia produced in rats by electroconvulsive shock. J. Comp. Physiol. Psychol. 60, 136-140.

Lice, C., Bailey, H., and Windle, W. 1950. J. Comp. Neurol. 92, 169.

Lichtman, C., and Hunt, R. 1971. Personality and organizational theory: A review of some conceptual literature. Psychol. Bull. 76, 271-294.

Lima, I., Luiz, R., and Carlini, E. 1966. Effects of strychnine and amphetamine on ribonucleic acid content in brain and liver of rats. Med. Pharmacol. Exp. 15, 1-6.

Lippe, W.R. 1976. Innate and experiental factors in the development of visual system: historical basis of current controversy, in Neural and Behavioral Specificity Gottleib, G. ed., New York Acad. Stud. Dev. Behav. Nerv. Syst. 3, 5-24.

Locke, J. 1690. An Essay Concerning Human Understanding. T. Basset, London. Recent edition by A.S. Pringle-Pattison, Clarendon Press, London, England.

Lorenz, K.Z. 1952. King Solomon's Ring. Cromwell, New York.

Louch, C.D., and Higginbotham, M. 1967. The relation between social rank and plasma corticosterone levels in mice. Gen. Comp. Endocrinol. 8, 441-444.

Lubar, J., Schortal, C., and Perachino, A. 1967. Non-visual functions of visual cortex in the cat. Physiol. Behav. 2, 179-184.

Lucero, M.A. 1970. Lengthening of R.E.M. sleep duration consecutive to learning in the rat. Brain Res. 20, 319-322.

Ludwig, A.M. 1972. Psychedelic effects produced by sensory overload. Am. J. Psychiat. 128, 114-117.

Ludwig, A.M., and Stark, L.H. 1973. Schizophrenia, sensory deprivation and sensory overload. J. Nerv. Ment. Dis. 157, 210-216.

Luria, A.R., Naydin, V.L., Tsvethova, L.S., and Vinarskaya, E.N. 1969. Restoration of higher cortical function following local brain damage, in Handbook of Clinical Neurology. Vol. 3, Vinken, P.J., and Bruyer, G. W., eds., Nath-Holland Publishing co., Amsterdam, 368-433.

Luxenburg, J. 1899. Neurol. Zentr. 8, 629.

Lynch, G., Smith, R., and Cotman, C.W. 1975. Recovery of function following brain damage: A consideration of some neural mechanisms, in The Neurophysiological Basis of Rehabilitative Medicine, Tobias, A., and Buerger, A. eds., Thomas, Springfield, Illinois.

Lynch, G., Stanfield, B., Parks, T., and Cotman, C.W. 1974. Evidence for selective postlesion growth in the dentate gyrus of the rat. Brain Res. 69, 1-11.

Lynn, R. 1966. Attention, Arousal and the Orientation Reaction. Pergamon Press, Oxford.

Maas, J.W. 1962. Neurochemical differences between two strains of mice. Science 137, 621-622.

Maier, S., Seligman, M., and Solomon, R. 1969. Pavlovian fear conditioning and learned helplessness: Effects on escape and avoidance behavior of (a) CS-US contingency and (b) the independence of the US and voluntary responding, in Punishment, Campbell, B. and Church, R., eds., Appleton-Century-Crofts, New York.

Mailloux, J.G., Edwards, H.P., Barry, W.F., Rowsell, H., and Ackhorn, E. 1974. Effects of differential rearing on cortical evoked potentials of the albino rat. J. Comp. Physiol. Psychol. 87, 475-480.

Maki, W.S. 1971. Failure to replicate effect of visual pattern restriction on brain and behavior. Nature (London) 233, 63-64.

Malacarne, M.W. 1785. Cited by Rosenzweig, M.R., Bennett, and Diamond, M.C. 1971. Chemical and Anatomical Plasticity of brain: Replication and Extensions, 1970, in Macromolecules and Behaviour (2nd ed.) Gaito, J. ed., Appleton-Century-Crofts, New York, 205-278.

Malkasian, D.R., and Diamond, M.C. 1971. The effects of environmental manipulation on the morphology of the neonate rat brain. Int. J. Neurosci. 2, 161-170.

Malmo, R.B. 1959. Activation: A neuropsychological dimension. Psychol. Rev. 66, 367-386.

Mann, S.A. 1925. Blood sugar studies in mental disorders. J. Ment. Sci. 71, 443-473.

Marcucci, F., Mussini, E., Valzelli, L., and Garattini, S. 1968. Decreases in N-acetyl-L-aspartic acid in brain of aggressive mice. J. Neurochem. 15, 53-54.

Maslow, A. 1971. The Farther Reaches of Human Nature. Viking, New York.

Mason, W.A. 1963. Social development of rhesus monkeys with restricted social experience. Percept. Mot. Skills 16, 263-270.

Mathies, H., Lossner, B., Ott, T., Pohle, W., and Ranca, C. 1973. The intraneuronal regulation of neuronal conectivity. In Pharmacology and the Future of Man. Proc. Int. Congr. Pharmacol. Fifth. 4, 29-38, Karger, Basel.

Maturana, H. 1960. The fine anatomy of the optic nerve of Anurans - An electron microscopic study. J. Biophys. Biochem. Cytol. 7, 107-120.

McCaman, R.E., and Hunt, J.M. 1965. Microdetermination of choline acetylase in nervous tissure. J. Neurochem. 12, 253-259.

McClelland, D., Ackinson, J., Clark, R., and Lowell, E. 1953. The Achievement Motive. Appleton-Century-Crofts, New York.

McGinty, D.J. 1971. Encephalization and the neural control of sleep. In Brain Development and Behavior, Sterman, M.B., McGinty, D.J., and Adinolfi, A.M.,eds., Academic Press, New York, 325-357.

McIver, A., Jeffrey, W., Stevenson, M., and Milson, H. 1968. Infantile handling and body temperature change and its implications for later responses. Trans. Acad. Sci. (St. Louis) 30, 992-1000.

Meier, G.W. 1968. The half told tale and the half replication. Dev. Psychobiol. 1, 77-78.

Meisami, E. and Timiras, P.S. 1971. Influence of early visual input on development of brain excitability in the rat. Am. J. Physiol. 220, 233-238.

Melzack, R. 1968. Early experience: a neuropsychological approach to hereditary-environment interactions, in Early Experience and Behavior. Newton, C., and Levine, S. eds., C.C. Thomas, Springfield, Illinois, 65-82.

Melzack, R. 1969. The role of early experience in emotional arousal. Ann. N.Y. Acad. Sci. 159, 721-730.

Merrill, E.G., and Wall, P.D. 1972. Factors forming the edge of a receptive field: The presence of relatively ineffective afferent terminals. J. Physiol. 226, 825-846.

Michael, R., and Gibbons, G. 1963. Interelationship between the endocrine system and neuropsychiatry, in International Review of Neurobiology, Pfieffer, C.C., and Smythies, J.R. eds., Academic Press, New York.

Michelson, W. 1970. Man and His Urban Environment: A Sociological Approach. Addison Wesley, Reading, Massachusetts.

Michelson, W. 1971. Some like it hot: Social participation and environmental use as functions of the season. Am. J. Sociol. 76, 1072-1083.

Miller, J.G. 1975. General systems theory, in Comprehensive Textbook of Psychiatry, Freedman, A., Kaplan, H., and Sadock, B.J. eds., Williams and Wilkins, Baltimore, Maryland, 75-88.

Milner, B. 1973. Hemispheric specialization. Scope and limits, in The Neurosciences 3rd Study Program, Schmitt, F.O., and Warden, F.G. eds., MIT Press, Cambridge, Massachusetts, 75-89.

Mischel, W. 1968a. Psychosocial deprivation and self control, in Perspectives On Human Deprivation. U.S. Department of Health, Education and Welfare, Washington, D.C.

Mischel, W. 1968b. Personality and Assessment. Wiley, New York.

Mitchell, D.E., Cynader, M., and Movshon, J.A. 1977. Recovery from the effects of monocular deprivation in kittens. J. Comp. Neurol. 176, 23-52.

Moliner, E. Ramon 1958. A study on neuroglia - the problem of transitional forms. J. Comp. Neurol. 110, 157-172.

Mollgaard, K., Diamond, M.C., Bennett, E.L., Rosenzweig, M.R., and Lindner, B. 1971. Quantitative synaptic changes with differential experience in rat brain. Int. J. Neurosci. 2, 113-128.

Moore, K.E., Sawdy, L.C., and Shaul, S.R. 1965. Effects of d-amphetamine on blood glucose and tissue glycogen levels of isolated and aggregated mice. Biochem. Pharmacol. 14, 197-204.

Moore, R.Y. 1974. Central regeneration and the recovery of function: The problem of collateral reinnervation, in Plasticity and the Recovery of Function in the Central Nervous System, Stein, D.C., Rosen, J.J., and Butters, N. eds., Academic Press, New York, 111-128.

Moore, R.Y. 1976. Synaptogenesis and the morphology of learning and memory, in Neural Mechanisms of Learning and Memory, Rosenzweig, M. R., and Bennett, E.L. eds., MIT Press, Cambridge, Massachusetts, 340-347.

Moos, R.H. 1973. Conceptualizations of human environments. Am. Psychol. 28, 652-665.

Moos, R.H. 1974. Evaluating Treatment Environments: A Social Ecological Approach. Wiley Interscience, New York.

Moos, R.H. 1975. Evaluating Correctional and Community Settings. Wiley Interscience, New York.

Moos, R. 1976. The Human Context: Environmental Determinants of Behavior. Wiley Interscience, New York.

Moos, R.H., and Insel, P.M. 1974. Issues in Social Psychology: Human Milieus. National Press Books, California.

Mori, S., and Shimizu, N. 1964. Electron microscopic histochemistry of cholinesterases in the rat brain. Histochemie 4, 65.

Movshon, J.A., and Blakemore, C. 1975. Functional cordotomies on the motilities of the lower limbs. Brain 96, 471-494.

Murphy, G., and Spohn, E. 1968. Encounter with Reality. Houghton Mifflin, Boston, Massachusetts.

Mushynski, W.E., Levitan, I.B., and Ramirez, G. 1973. Competition hybridization studies on brain ribonucleic acid from rats reared in enriched and impoverished environments. J. Neurochem. 20, 309-317.

Nair, V., Brown, T., Bau, D., and Siegel, S. 1969. Evidence for hypothalamic regulation of hepatic hexobarbital oxidase in rats. Pharmacologist 11, 121.

Nathan, P., and Smith, M. 1973. Effects of 2 unilateral cordotomies on the motilities of the lower limbs. Brain 96, 471-494.

Needham, J. 1956. Science and Civilization in China. Cambridge Univ. Press, Cambridge, England.

Nelson, K. 1973. Structure and strategy in learning to talk. Monogr. Soc. Res. Child Dev. 38, (1-2, Serial No. 149).

Oppenheimer, J.R. 1954. Science and the Common Understanding. Oxford University Press, New York.

Orkand, R., Nikols, J., and Kuffler, S. 1966. The effect of nerve impulses on the membrane potential of glial cells in the central nervous system of amphibia. J. Neurophysiol. 29, 788-806.

Oswald, I. 1969. Human brain protein, drugs and dreams. Nature (London) 233, 893-897.

Parker, D.C., Sassin, J.F., Mace, J.W., Gotlin, R.W., and Rossman, L.G. 1969. Human growth hormone release during sleep: Electroencephalographic correlation. J. Clin. Endocrinol. 29, 871-874.

Pauling, L. 1968. Orthomolecular psychiatry. Science 160, 265.

Perlia, R. 1889. Uber ein neues Opticus Centrum beim, Hubne, Arch. Ophthamol. 35, 20-24.

Peter, L.J., and Hull, R. 1969. The Peter Principle. William Morrow, New York.

Peters, A. 1960. The formation and structure of the myelin sheath in the central nervous system. J. Biophys. Biochem. Cytol. 8, 431-446.

Peterson, R., and Kernell, D. 1970. Effects of nerve stimulation on the metabolism of ribonucleic acid in the molluscan giant neuron. J. Neurochem. 17, 1075-1085.

Pettigrew, J., Olson, C., and Barlow, H. 1973. Kitten visual cortex: short term, stimulus-induced changes in connectivity. Science 180, 1202-1203.

Pevzner, L.Z. 1966. Nucleic acid changes during behavioral events, in Macromolecules and Behaviour, Gaito, J. ed. Appleton-Century-Crofts, New York.

Pfaffman, C. 1960. The pleasures of sensation, Psychol. Rev. 67, 253-268.

Prescott, J.W. 1970a. Early somatosensory deprivation as an ontogenetic process in abnormal development of the brain and behavior. Medical Primatology, Proc. 2nd Conf. Exp. Med. Surg. Primates, New York, 1969, Karger, Basel.

Prescott, J.W. 1970b. A developmental neural-behavioral theory of socialization: Maternal social deprivation as functional somatosensory differentation in the abnormal development of the brain and behavior. Am. Psychol. Assoc. Symp., Florida, September 8.

Prescott, J.W. 1971. Sensory deprivation versus sensory stimulation during early development: a comment on Berkowitz's study. J. Psychol. 77, 189-191.

Prestige, M.C. 1970. Deafferentation, degeneration, and the role of the periphery: quantitative considerations, in The Neurosciences Second Study Program, Schmitt, F.O. ed. Rockefeller University Press, New York, 73-82.

Pribram, K., and Gill, N. 1976. Freud's Neuropsychology: Preface to Contemporary Cognitive Theory and Biological Psychiatry. Basic Books, New York.

Proshansky, H., Ittelson, W., and Rielin, L. eds. 1970. Environmental Psychology: Man and His Physical Settings. Holt, Rinehart and Winston, New York.

Pryor, G.T., Otis, L.S., Scott, M.K., and Colwell, J.J. 1967. Duration of chronic electroshock treatment in relation to brain weight, brain chemistry and behavior. J. Comp. Physiol. Psychol. 63, 236-239.

Pugh, D. 1966. Modern organization theory: A psychological and sociological study. Psychol. Bull. 66, 235-251.

Putnam, H. 1964. A philosopher looks at quantum mechanics, in Philosophy of Mathematics, Benacerrof, P., and Putnam, H. eds. Prentice-Hall, Englewood Cliffs, New Jersey.

Quamina, A., and Bogoch, S. 1966. In Proc. XIII Colloq. Protides of Biological Fluids, Peeters, H. ed. Brugge, Elsevier, Amsterdam, 211.

Quay, W.B., Bennett, E.L., Rosenzweig, M.R., and Krech, D. 1969. Effects of isolation and environmental complexity on brain and pineal organ. Physiol. Behav. 4, 489-494.

Radhakrishnan, S. 1958. Indian Philosophy. Macmillan, New York.

Rajneesh, B.S. 1975a. The Way of the White Cloud. Rajneesh Foundation, Poona, India.

Rajneesh, B.S. 1975b. Just Like That. Rajneesh Foundation, Poona, India.

Ram Dass. 1975. Association Transpers. Psychology Newsletter, Winter.

Ram Dass. 1978. Journey of Awakening: A Meditator's Guidebook. Bantam, New York.

Ramirez, G., Karlsson, B., and Perrin, C.L. 1974. Effects of environmental complexity on glucosamine incorporation into rat cerebral cortex and hippocampus *in vivo*. Brain Res. 79, 296-300.

Ramon y Cajal, S. 1895. Les Nouvelles idees sur la structure de systeme merveux chez l'homme et chez les vertebres. Edition francaise revue et augmentee par l'auteur. Reinwals, Paris, France, 78.

Randrup, A., and Munkvad, I. 1968. Behavioral stereotypies induced by pharmacological agents. Pharmakopsychiatr. Neuro-psychopharmakol. 1, 18-26.

Randrup, A., and Munkvad, I. 1970. Biochemical, anatomical and psychological investigations of stereotyped behavior induced by amphetamine, in Symposium on Amphetamine and Related Drugs, Costa, E. and Garattini, S. eds. Raven Press, New York, 695-713.

Reis, D., and Fuxe, K. 1969. Brain norepinephrine: Evidence that neuronal release is essential for sham rage behavior following brainstem transection in the cat. Proc. Nat. Acad. Sci. USA. 64, 108-112.

Riege, W.H. 1971. Environmental influences on brain and behavior of old rats. Dev. Psychol. 4, (a).

Riege, W.H., and Morimoto, H. 1970. Effects of chronic stress and differential environments upon brain weight and biogenic amine levels in rats. J. Comp. Physiol. Psychol. 71, 396-404.
Richardson, S. 1976. The influence of severe malnutrition in infancy on the intelligence of children at school age: An ecological perspective. In Environments as Therapy for Brain Dysfunction, Walsh, R., and Greenough, W. eds. Plenum Press, New York, 256-275.
Riesen, A.H. 1961. Stimulation as a requirement for growth and function in behavioral development, in Functions of Varied Experience, Fiske, D.W., and Maddi, S.R. eds. Dorsey Press, Homewood, 57-80.
Riesen, A.H. 1966. Sensory deprivation. Prog. Physiol. Psych. 1, 117-147.
Rizzolatti, F., and Tradari, V. 1971. Pattern discrimination in monocularly reared cats. Exp. Neurol. 33, 181-194.
Roberts, K. 1970. On looking at an elephant: An evaluation of cross cultural research related to organizations. Psychol. Bull. 74, 327-350.
Roberts, R.C. 1967. Can. J. Genet. Cytol. 9, 419.
Rose, S.P. 1967. Changes in visual cortex on first exposure of rats to light: effect on incorporation of tritiated lysine into protein. Nature (London) 215, 253-255.
Rose, S.P., and Sinha, A.K. 1974. Incorporation of $H^3$ lysine into a rapidly labelling neuronal protein fraction in visual cortex is suppressed in dark reared rats. Life Sci. 15, 223-230.
Rosenzweig, M.R. 1964. Effects of heredity and environment on chemistry, brain anatomy and learning ability in the rat. Symposium on Physiological Determinants of Behavior: Implications for Mental Retardation. Kansas Study Educ. 14, 3-34.
Rosenzweig, M.R. 1966. Environmental complexity, cerebral change, and behavior. Am. J. Psychol. 21, 321-332.
Rosenzweig, M.R. 1967. Effects of experience on brain chemistry and brain anatomy. Accad. Naz. Lincei. Roma 109, 43-63.
Rosenzweig, M.R. 1969. Mechanisms by which environmental complexity may alter brain chemistry and anatomy. Paper delivered at Int. Cong. Psychol. London, July.
Rosenzweig, M.R. 1978. Personal communication.
Rosenzweig, M.R., and Bennett, E.L. 1968. Drugs modulate effects of environment on brain growth. Proceedings 76th Ann. Conv. Amer. Psychol. Assoc. 3, 269-270.
Rosenzweig, M.R., and Bennett, E.L. 1969. Effects of differential environments on brain weights and enzyme activities in gerbils, rats and mice. Dev. Psychobiol. 2, 87-95.
Rosenzweig, M.R., and Bennett, E.L. 1972. Cerebral changes in rats exposed individually to an enriched environment. J. Comp. Physiol. Psychol. 80, 304-313.
Rosenzweig, M.R., and Bennett, E.L. 1976. Enriched environments: Facts, factors, and fantasies, in Knowing, Thinking, and Believing, Petrinovich, L., and McGaugh, J. eds. Plenum Press, New York.
Rosenzweig, M.R., and Bennett, E.L. 1977. Effects of environmental enrichment or impoverishment on learning, and on brain values in rodents, in Genetics, Environment, and Intelligence, Aliverio, A. ed., Elsevier, Holland, 163-196.
Rosenzweig, M.R., and Bennett, E.L. 1978. Experiential influences on brain anatomy and chemistry in rodents, in Studies on the Development of Behavior and the Nervous System, Vol. 4. Early Influences, Gottleib, G. ed., Academic Press, New York, 289-327.
Rosenzweig, M.R., Bennett, E.L., and Diamond, M.C. 1967. Effects of differential environments on brain anatomy and brain chemistry. Proc. Amer. Psychopatholog. Assoc. 56, 45-56.

Rosenzweig, M.R., Bennett, E.L., and Diamond, M.C. 1968. Modifying brain chemistry and anatomy by enrichment or impoverishment of experience, in Early Experience and Behavior: The Psychobiology of Development, Newton, G., and Levine, S. eds. C.C. Thomas, Springfield, Illinois, 258-298.

Rosenzweig, M.R., Bennett, E.L. and Diamond, M.C. 1971. Chemical and anatomical plasticity of brain: Replications and extensions, in Macromolecules and Behavior (2nd ed.), Gaito, J. ed. Appleton-Century-Crofts, New York, 205-278.

Rosenzweig, M.R., Bennett, E.L., and Diamond, M.C. 1972. Cerebral effects of differential environments occur in hypophysectomized rats. J. Comp. Physiol. Psychol. 79, 56-66.

Rosenzweig, M.R., Bennett, E.L., Diamond, M.C., Wu, S., Slagle, R.W., and Saffran, E. 1969. Influences of environmental complexity and visual stimulation on development of occipital cortex in rats. Brain Res. 14, 427-445.

Rosenzweig, M.R., Krech, D., and Bennett, E.L. 1958. Brain chemistry and adaptive behavior, in Biological and Biochemical Bases of Behavior. Harlow, H.F., and Woolsey, C.N. eds. Univ. of Wisconsin Press, Madison, Wisconsin, 367-400.

Rosenzweig, M.R., Krech, D., and Bennett, E.L. 1960. A search for relations between brain chemistry and behavior. Psychol. Bull. 57, 476-492.

Rosenzweig, M.R., Krech, D., Bennett, E.L., and Diamond, M.C. 1962. Effects of environmental complexity and training on brain chemistry and anatomy. J. Comp. Physiol. Psychol. 55, 429-437.

Rosenzweig, M.R., Mollgaard, K., Diamond, M.C., and Bennett, E.L. 1972. Negative as well as positive synaptic changes may store memory. Psychol. Rev. 79, 93-96.

Rosenzweig, N.R. 1959. Sensory deprivation and schizophrenia: Some clinical and theoretical similarities. Am. J. Psychiat. 116, 326-329.

Rothgeb, C.L. 1971. Abstracts of the Standard Edition of the Complete Psychological Works of Sigmund Freud. U.S. Department of Health, Education and Welfare, Maryland.

Ruiz-Marcos, A., and Valverde, F. 1969. The temporal evolution of the distribution of dendritic spines in the visual cortex of normal and dark-raised mice. Exp. Brain Res. 8, 284-294.

Ruiz-Marcos, A., and Valverde, F. 1970. Dynamic architecture of the visual cortex. Brain Res. 19, 25-39.

Russell, P.A. 1971. Infantile stimulation in rodents: A consideration of possible mechanisms. Psychol. Bull. 75, 192-202.

Russet, B. et al. 1964. World Handbook of Political and Social Indicators. Yale Univ. Press, New Haven.

Rutledge, L.T. 1976. Synaptogenesis: Effects of synaptic use, in Neural Mechanisms of Learning and Memory, Rosenzweig, M.R., and Bennett, E.L. eds. MIT Press, Cambridge, Massachusetts.

Rutledge, L., Duncan, J., and Beatty, N. 1969. A study of pyramidal cell axon collaterals in intact and partially isolated adult cerebral cortex. Brain Res. 16, 15-22.

Rychlak, J.F. 1968. A Philosophy of Science for Personality Theory. Houghton Mifflin, Boston.

Ryugo, R., Globus, A., and Killachey, H.B. 1975. Increased spine density in auditory cortex following visual or somatic deafferentation. Brain Res. 90, 143-146.

Sackett, G.P. 1972. Exploratory behavior of rhesus monkeys as a function of rearing experiences and sex. Dev. Psychol. 6, 260-270.

Sacks, W. 1973. Disorders of glucose metabolism in brain dysfunction, in Biology of Brain Dysfunction. Vol. 1, Gaull, G.E. ed. Plenum Press, New York, 143-189.

Sahakian, B., Rolobins, T., Morgan, M., and Iversen, S. 1975. The effects of psychomotor stimulants on stereotypy and locomotor activity in socially deprived and control rats. Brain Res. 84, 195-205.

Sargent, F. 1972. Man-environment: Problems for public health. Am. J. Public Health 62, 628-633.

Sarno, M.T. 1970. Speech therapy and language recovery in severe aphasia. J. Speech Hear. Res. 13, 607-625.

Schain, R.J., and Watanabe, K. 1976. Origin of brain growth retardation in young rats treated with phenobarbital. Exp. Neurol. 50, 806.

Shapiro, S., and Vukovich, K.R. 1970. Early experience effects upon cortical dendrites: A proposed model for development. Science 167, 292-294.

Schaeffer, T. 1968. Some methodological implications of the research on early handling in the rat, in Early Experience and Behavior, Newton, G., and Levine, S. eds. C.C. Thomas, Springfield, Illinois, 102-141.

Scherrer, J., and Fourment, A., 1964. Electro-cortical effects of sensory deprivation during development, in Progress in Brain Research Vol. 9, Himwich, W., and Himwich, H. eds. Elselvier, Amsterdam, 103-112.

Schlosberg, H. 1954. Three dimensions of emotion. Psychol. Rev. 61, 81-88.

Schmidt, H.J. 1976. A theory of the serial lesion effect. Brain Res. Bull. 1, 261-262.

Schroedinger, E. 1964. My View of the World. Cambridge Univ. Press, London, England, 62.

Schroedinger, E. 1969. What is Life? and Mind and Matter. Cambridge Univ. Press, London, England, 137.

Scott, J.P. 1947. "Emotional" behavior of fighting mice caused by conflict between weak stimulatory and weak inhibitory training. J. Comp. Physiol. Psychol. 40, 275-282.

Scott, J.P., and Fredericson, E. 1951. The cause of fighting in rats and mice. Physiol. Zool. 24, 273-309.

Scriven, M. 1969. Psychology without a paradigm, in Clinical Cognitive Psychology, Berger, L. ed. Prentice-Hall, Englewood Cliffs, New Jersey.

Senf, G. 1976. Some methodological considerations in the study of abnormal conditions, in Environments as Therapy for Brain Dysfunction, Walsh, R., and Greenough, W. eds. Plenum Press, New York, 343-360.

Sengtsan. 1975. Verses on the Faith Mind, Clarke, R.B., trans. Zen Centre, Sharon Springs, New York.

Shapiro, D., and Walsh, R.N. eds. 1978. The Science of Meditation: Research and Theory. Aldine Press, New York (In press).

Sharpless, S., and Halpern, L. 1962. The electrical excitability of chronically isolated cortex studies by means of permanently implanted electrodes. Electroencephalogr. Clin. Neurophysiol. 14, 244-255.

Sharpless, S.K. 1970. Isolated and deafferented neurons: Disuse supersensitivity, in Basic Mechanisms of the Epilepsies, Jasper, H.H. ed. Little, Brown, Boston.

Sherlock, D., Cummins, R.A., Walsh, R.N., and Budtz-Olsen, O.E. 1973. Reversibility of deficits in problem solving ability following early environmental deprivation. Proc. Aust. Soc. Med. Res. 3, 174.

Sherman, S.M. 1974. Permanence of visual perimetry deficits in monocularly and binocularly deprived cats. Brain Res. 73, 401-501.

Sherman, S.M., and Sanderson, K.J. 1972. Binocular interaction on cells of the dorsal lateral geniculate nucleus of visually deprived cats. Brain Res. 37, 126-131.

Shilyagina, N.N. 1973. Electrical activity of the cortical and subcortical parts of the rabbit brain following early visual deprivation. J. Higher Nervous Act. (Russ.) 23, 1066-1073.

Sholl, D.A. 1956. The Organization of the Cerebral Cortex. Methuen, London.

Shopsin, B., Stern, S., and Gershon, S. 1972. Altered carbohydrate metabolism during treatment with lithium carbonate. Arch. Gen. Psychiat. 26, 566-571.

Singh, D., Johnston, R., and Klosterman, H. 1967. Effect on brain enzyme and behavior in the rat of visual pattern restriction in early life. Nature (London) 216, 1337-1338.

Sloviter, H.A., and Kamimoto, T. 1970. The isolated perfused rat brain metabolizes mannose but not maltose. J. Neurochem. 17, 1109-1112.

Smart, I., and Leblond, C.P. 1961. Evidence for division and transformation of neuroglia cells in the mouse brain, as derived from radioautography after injection of thymidine-$H^3$. J. Comp. Neurol. 116, 349-368.

Soemmering, S.T. 1971. Von Bauedes Menscheichen Koerpus. Vol. 5, part 1. Barrentropp and Wenner, Frankfeut Am Main.

Sokolov, E.N. 1960. Neuronal models and the orienting reflex, in The Central Nervous System and Behavior, Brazier, M.A. ed. G. Macey, New York.

Solkoff, N., Yaffir, S., Weintraub, D., and Blase, B. 1969. Effects of handling on the subsequent development of premature infants. Dev. Psychol. 1, 765-768.

Sollberger, A. 1960. Studies of temporal variations in biological variates. Supplement Rept., 5th Int. Conf. Soc. Biol. Rhythm. Stockholm, 111 pp.

Sollberger, A. 1965. Biological Rhythm Research. Elsevier, Amsterdam.

Sotelo, C., and Palay, S.L. 1971. Altered axons and axon terminals in the lateral vestibular nucleus of the rat. Lab. Invest. 25, 653-671.

Sperry, R.W. 1968. Plasticity of neural maturation. The Society for Developmental Biology, Symposium, 27th. New York, June, 1963. "The Emergence of Order in Developing Systems." Locke, M. ed. Academic Press, New York and London.

Sperry, R.W. 1977. Bridging science and values: A unifying view of mind and brain. Am. Psychol. 32, 237-245.

Spoerlein, M.T. 1968. Pharmacologist 10, 172. Cited by Davis and Brister, 1971.

Spurzheim, J.G. 1815. The Physiognomical System of Drs. Gall and Spurzheim. Baldwin, Cradock, and Joy, London. 554-555.

Stapp, H.P. 1971. S-matrix interpretation of quantum theory. Phys. Rev. D3, 1303-1320.

Stern, P., McDowell, F., and Robinson, J.M. 1971. Factors influencing stroke rehabilitation. Stroke 2, 213-218.

Suomi, S.J., and Harlow, H.F. 1972. Social rehabilitation of isolate-reared monkeys. Dev. Psychol. 6, 487-496.

Suzuki, D.T. 1952. Studies in the Lankavatara Sutra. Routledge and Kegan Paul, London.

Suzuki, D.T. 1959. Preface to Mahayana Buddhism, B. L. Suzuki. Allen and Unwin, London.

Suzuki, D.T. 1968. The Essence of Buddhism. Hozokan, Kyoto.

Szeligo, F., and Leblond, P. 1977. Response of the three main types of glial cells of cortex and corpus callosum in rats handled during suckling or exposed to enriched, control, and impoverished environments during weaning. J. Comp. Neurol. 172, 247-264.

Tagney, J. 1973. Sleep patterns related to rearing rats in enriched and impoverished environments. Brain Res. 53, 353-361.

Tapp, J., and Markowitz, H. 1963. Infant handling: Effects on avoidance learning, brain weight and cholinsterase activity. Science 140, 486-487.

Tart, C. 1975a. States of Consciousness. E.P. Dutton, New York.

Tart, C. ed. 1975b. The Transpersonal Psychologies. Harper and Row, New York.

Terenius, L., and Wahlstrom, A. 1975. Morphine-like ligand for opiate receptors in human CSF. Life Sci. 16, 1759-1764.

Thera, N. 1962. The Heart of Buddhist Meditation. Samuel Weiser, New York.

Thetford, W., Schucman, H., and Walsh, R. 1978. Other Psychological Theories, in. A Comprehensive Textbook of Psychiatry 3rd ed., Kaplan, H., Friedman, A.,and Sadock, B. eds. Williams and Wilkins, Baltimore. (In press).

Thierry, A.M., Javoy, F., Glowinski, J., and Kety, S.S. 1968. Effects of stress on the metabolism of norepinephrine, dopamine and serotonin in the central nervous system of the rat. I. Modifications of norepinephrine turnover. J. Pharmacol. Exp. Ther. 163, 163-171.

Thiessen, D., Zolman, J., and Rodgers, D. 1962. Relation between adrenal weight, brain cholinesterase activity, and hole-in-the-wall behavior of mice under different living conditions. J. Comp. Physiol. Psychol. 55, 186-190.

Thoman, E., and Levine, S. 1969. Role of maternal disturbance and temperature change in early experience studies. Physiol. Behav. 4, 143-145.

Thompson, E.R., and Grusec, G. 1970. Studies of early experience, in Carmichael's Manual of Child Psychology Vol. 1, (3rd ed.), Mussen, P. H. ed. Wiley, New York.

Thoresen, C.E., and Mahoney, M.J. 1974. Behavioral Self-Control. Holt, Rinehart and Winston, New York.

Tilopa 1977. Cited by J. Goldstein in a talk given at Barre, Massachussetts, November.

Timiras, P. 1972. Oestrogens as organizers of CNS functions, in The Influence of Hormones on the Nervous System, Ford, D., ed. International Society of Psychoneuroendocrinology, Karger, Basel, Switzerland.

Torack, R.M., and Barnett, R.J. 1962. Fine structural localization of cholinesterase activity in the rat brain stem. Exp. Neurol. 6, 224.

Tucek, S., and Cheng, S.C. 1970. Precursors of acetyl groups in acetylcholine in the brain *in vivo*. Biochim. Biophys. Acta. 208, 538-540.

Uphouse, L. 1978. *In vitro* RNA synthesis by chromatin from three brain regions of differentially reared rats. Behav. Biol. 22, 39-50.

Uphouse, L.L., and Bonner, J. 1975. Preliminary evidence for the effects of environmental complexity on hybridization of rat brain RNA to rat unique DNA. Dev. Psychobiol. 8, 171-178.

Uphouse, L.L., MacInnes, J., and Schlesinger, K. 1974. Role of RNA and protein in memory storage: A review. Behav. Genet. 4, 29-82.

Uphouse, L., and Moore, R. 1978. Effect of rearing condition on *in vitro* RNA synthesis by brain chromatin. Behav. Biol. 22, 23-39.

Valverde, F. 1967. Apical dendritic spines of the visual cortex and light deprivation in the mouse. Exp. Brain. Res. 3, 337-352.

Valverde, F. 1968. Structural changes in the area striata of the mouse after enucleation. Exp. Brain. Res. 5, 274-292.

Valverde, F. 1971. Rate and extent of recovery from dark rearing in the visual cortex of the mouse. Brain. Res. 33, 1-11.

Valverde, F., and Ruiz-Marcos, A. 1969. Dendritic spines in the visual cortex of the mouse. Introduction to a mathematical model. Exp. Brain. Res. 8, 269-283.

Valzelli, L. 1967a. Drugs and aggressiveness. Adv. Pharmacol. 5, 79-108.

Valzelli, L. 1967b. Biological and pharmacological aspects of aggressiveness in mice, in Neuropsychopharmacology, Proc. 5th Int. CINP Congr., Brill, H. ed. Excerpta Medica Fdn., Amsterdam, 28-31.

Valzelli, L. 1971. Agressivité chez le rat et la souris: Aspects compartementaux et biochimiques. Actual. Pharmacol. 24, 133-152.

Valzelli, L. 1973a. The "isolation syndrome" in mice. Psychopharmacologia 31, 305-320.

Valzelli, L. 1973b. Psychopharmacology: An Introduction to Experimental and Clinical Principles. Spectrum Publ., New York.

Valzelli, L. 1973c. Activity of benzodiazepines on aggressive behavior in rats and mice, in The Benzodiazepines, Garattini, S, Mussini, E., and Randall, L.O. eds. Raven Press, New York, 405-417.

Valzelli, L., and Bernasconi, S. 1971. Differential activity of some psychotropic drugs as a function of emotional level in animals. Psychopharmacologia 20, 91-96.

Valzelli, L., and Garattini, S. 1968. Behavioral changes and 5-HT turnover in animals. Adv. Pharmacol. 6B, 249-260.

Valzelli, L., Giacalone, E., and Garattini, S. 1967. Pharmacological control of aggressive behavior in mice. Eur. J. Pharmacol. 2, 144-146.

Van Harreveld, A., and Fifkova, E. 1975a. Rapid freezing of deep cerebral structures for electron microscopy. Anat. Rec. 182, 377-386.

Van Harreveld, A., and Fifkova, E. 1975b. Swelling of dendritic spines in the fascia dentata after stimulation of the perforant fibers as a mechanism of post-tetanic potentiation. Exp. Neurol. 49, 736-749.

Van Harreveld, A., and Khattab, F.I. 1967. Changes in cortical extracellular space during spreading depression investigated with the electron microscope. J. Neurophysiol. 30, 911-929.

Varela, F. 1976. On observing natural systems. Coevolution Summer, 26-31.

Vedernikov, Y.P. 1970. Psychopharmacologia 17, 283-288. Cited by Davis and Brister, 1971.

Vivekananda, S. 1972. Jnana Yoga. Ramakrishman-Vivekananda Center, New York.

Vogt, M. 1960. Epinephrine and Norepinephrine, in Hormones in Human Plasma, Nature and Transport, Antoniades, H.N. ed. Little Brown, Boston.

Volkmar, F.R., and Greenough, W.T. 1972. Rearing complexity affects branching of dendrites in the visual cortex of the rat. Science 176, 1445-1447.

Von Bertalanffy, L. 1968. General Systems Theory. George Braziller, New York, 45.

Von Noorden, G.K., Crawford, M.L.J., and Middleditch, P.R. 1976. The effect of monocular visual deprivation: disuse or binocular interaction? Brain Res. 111, 277-285.

Vrba, R. 1962. Glucose metabolism in rat brain *in vivo*. Nature (London) 195, 663-665.

Vrensen, G., and DeGroot, D. 1974. The effect of dark rearing and its recovery on synaptic terminals in the visual cortex of rabbits: A quantitative electron microscopic study. Brain Res. 25, 263-278.

Vrensen, G., and DeGroot, D. 1977. Quantitative aspects of the synaptic organization of the superior colliculus in control and dark reared rabbits. Brain Res. 134, 417-429.

Walker, C.A. 1972. Twenty-four hour $LD_{50}$ rhythms of d,l-amphetamine in isolated and aggregated mice. In, Drug addiction: Experimental Pharmacology. Vol.1. Futura Publ. Co., Mount Kisco, New York, 247-253.

Wall, P.D., and Egger, M.D. 1971. Formation of new connections in adult rat brain after partial deafferentation. Nature (London) 232, 592-595.

Walsh, R.N. 1975. Beating the menstrual blues. Tex. Rep. Biol. Med. 32, 842.

Walsh, R. 1977. Initial meditative experiences: I. J. Transpers. Psychol. 9, 151-192.

Walsh, R. 1978. Initial meditative experiences: II. J. Transpers. Psychol. 10, 1-28.

Walsh, R.N., Budtz-Olsen, O.E., Penny, J.E., and Cummins, R.A. 1969. The effect of environmental complexity on the istology of the rat hippocampus. J. Comp. Neurol. 137, 361-366.

Walsh, R.N., Budtz-Olsen, O.E., Torok, A., and Cummins, R.A. 1971. Environmentally induced changes in the dimensions of the rat cerebrum. Dev. Psychobiol. 4, 115-122.

Walsh, R.N., and Cummins, R.A. 1975. Mechanisms mediating the production of environmentally induced brain changes. Psychol. Bull. 82, 986-1000.

Walsh, R.N., and Cummins, R.A. 1976a. Neural responses to therapeutic sensory environments, in Environments as Therepy for Brain Dysfunction, Walsh, R.N., and Greenough, W.T. eds. Plenum Press, New York, 171-200.

Walsh, R.N., and Cummins, R.A. 1976b. The open field test: A critical review. Psychol. Bull. 83, 482-504.

Walsh, R.N., and Cummins, R.A. 1976c. Effects of differential sensory environments on the electron microscopy of the rat occipital cortex. Proc. Soc. Neurosci. Ann. Meet. 2, 839.

Walsh, R.N., and Cummins, R.A. 1977. Old brains can. Paper presented at VI World Cong. Psychiatry. Honolulu, September.

Walsh, R.N., and Cummins, R.A. 1978a. Unpublished data.

Walsh, R.N., and Cummins, R.A. 1978b. Caveats for future research on the open field test. Psychol. Bull. 85, 587-589.

Walsh, R.N., and Cummins, R.A. 1979. Changes in hippocampal neuronal nuclei in response to environmental stimulation. Internat. J. Neurosci. 9, 209-212.

Walsh, R.N., Cummins, R.A., and Budtz-Olsen, O.E. 1973a. Environmentally induced changes in the dimensions of the rat cerebrum: A replication and extension. Dev. Psychobiol. 6, 3-8.

Walsh, R.N., Cummins, R.A., and Budtz-Olsen, O.E. 1973b. Cerebral plasticity is retained in old age. Tex. Rep. Biol. Med. 31, 603-604.

Walsh, R.N., Cummins, R.A., and Budtz-Olsen, O.E. 1974. The time course of environmentally induced changes in brain and body weight. Selec. Doc. Psychol. 5, 283-288.

Walsh, R.N., Cummins, R.A., Budtz-Olsen, O.E., O'Rourke, G., Brown, H., and Cameron, J. 1971b. Effects of environmental enrichment and deprivation on cerebral histology and composition. Proc. Aust. Soc. Med. Res. 2, 74.

Walsh, R.N., Cummins, R.A., Budtz-Olsen, O.E., and Torok, A. 1972. Effects of environmental enrichment and deprivation on rat frontal cortex. Int. J. Neurosci. 4, 239-242.

Walsh, R.N., and Greenough, W.T., eds. 1976b. Environments as therapeutic agents: an introduction, in Environments as Therapy for Brain Dysfunction, Walsh, R.N., and Greenough, W.T. eds. Plenum Press, New York, 1-9.

Walsh, R.N., and Greenough, W.T., eds. 1976b. Environments as Therapy for Brain Dysfunction. Plenum Press, New York.

Walsh, R., and Shapiro, D. 1978a. An historical and conceptual model of meditation research, in The Science of Meditation: Research and Theory, Shapiro, D., and Walsh, R., eds. Aldine, New York.

Walsh, R.N., and Shapiro, D. eds. 1978b. Beyond Health and Normality: Explorations of Extreme Psychological Wellbeing. Van Nostrand Reinhold, New York.

Walsh, R., and Vaughan, F. 1978. Comparative models of the person and psycholtherapy, in Transpersonal Psychotherapy, Boorstein, S., ed. Science and Behavior Books, Palo Alto, California (In press).

Walsh, R.N., and Vaughan, F. eds. 1978. Beyond Ego: Transpersonal Dimensions in Psychology. Tarcher Press, Los Angeles, California.

Ward, A., and Kennard, M. 1942. Effect of cholinergic drugs on recovery of function following lesions of the central nervous system in monkeys. Yale J. Biol. Med. 15, 189-228.

Watson, C., and Kennard, M. 1945. The effect of anticonvulsant drugs on recovery of function following cerebral cortical lesions. J. Neurophysiol. 8, 221-231.

Weaver, L.C., and Kerley, T.L. 1962. Stain difference in response of mice to d- amphetamine. J. Pharmacol. Exp. Ther. 135, 240-244.

Webb, W.B., and Friedman, J. 1971. Attempts to modify the sleep patterns of the rat. Physiol. Behav. 6, 459-460.

Weiss, J.M. 1968. Effects of coping responses on stress. J. Comp. Physiol. 65, 251-260.

Welch, B.L. 1965. Psychophysiological response to the mean level of environmental stimulation: A theory of environmental integration, in Symp. Med. Aspects of Stress in Military Climate. U.S. Govt. Printing Office, Washington, D.C., 39-96.

Welch, B.L., 1967. Discussion of aggression, defense, and neurohumors by A.B. Rothballer, in Aggression and Defense, Neural Mechanisms and Social Patterns: U.C.L.A. Forum in Medical Sciences, Vol. 7, Clemente, C.D., and Lindsley, D.B. eds. Univ. of Calif. Press, 150.

Welch, B.L., Brown, D.G., Welch, A.S., and Lin, D.C. 1974a. Isolation, restrictive confinement, or crowding of rats for one year. I. Weight, Nucleic acids and protein of brain regions. Brain Res. 75, 71-84.

Welch, B.L. Hendley, E.D., and Turek, I. 1974b. Norepinephrine uptake into cerebral cortical synaptosomes after one fight or electroconvulsive shock. Science 183, 220-221.

Welch, B.L. , and Welch, A.S. 1966. Graded effect of social stimulation upon d-amphetamine toxicity, aggressiveness and heart and adrenal weight. J. Pharmacol. Exp. Ther. 151, 331-338.

Welch, A.S., and Welch, B.L. 1968a. Effect of stress and parachlorophenylalanine upon brain serotonin, 5-hydroxyindoleacetic acid and catecholamines in grouped and isolated mice. Biochem. Pharmacol. 17, 699-708.

Welch, A.S., and Welch, B.L. 1968b. Reduction of norepinephrine in the lower brain stem by psychological stimulus. Proc. Nat. Acad. Sci. USA, 60, 478-481.

Welch, A.S., and Welch, B.L. 1968c. Greater lowering of brain and adrenal catecholamines in group-housed than in individually-housed mice administered DL-methyltyrosine. J. Pharm. Pharmacol. 20, 244-246.

Welch, A.S., and Welch, B.L. 1971. Isolation, reactivity and aggression: evidence for an involvement of brain catecholamines and serotonin, in The Physiology of Aggression and Defeat, Eleftheriou, B.E., and Scott, J.P. eds. Plenum Press, New York, 91-142.

Welch, B.L., and Welch, A.S., 1968d. Rapid modification of isolation-induced aggressive behavior and elevation of brain catecholamines and serotonin by the quick-acting monoamine oxidase inhibitor, pargyline. Commun. Behav. Biol. 1, 347-351.

Welch, B.L., and Welch, A.S. 1969a. Aggression and the biogenic amine neurohumors, in Aggressive Behavior, Garattini, S., and Sigg, E.B. eds. Excerpta Medica Fdn., Amsterdam, 188-202.

Welch, B.L., and Welch, A.S. 1969b. Sustained effects of brief daily stress (fighting) on brain and adrenal catecholamines, and adrenal, spleen, and heart weights of mice. Proc. Nat. Acad. Sci. USA 64, 100-107.

Welch, B.L., and Welch, A.S. 1973. Chronic social stimulation and tolerance to amphetamine: Interacting effects of amphetamine and natural nervous stimulation upon brain amines and behavior, in Current Concepts in Amphetamine Abuse, Ellinwood, E.H., Jr., and Cohen, S. eds. U.S. Govt. Printing Office, Washington, D.C., 107-115.

West, R., and Greenough, W.T. 1972. Effect of environmental complexity on cortical synapses of rats: preliminary results. Behav. Biol. 7, 279-284.

Westrum, L.E. 1975. Electron microscopy of synaptic structures in olfactory cortex of early postnatal rats. J. Neurocytol. 4, 713-732.

Wheeler, J.A. 1973. The Physicists' Conception of Nature, Mehra, J. ed. D. Reidel, Dordrecht, Holland, 244.

White, J.L., and LaBarba, R.L. 1976. The effects of tactile and kinesthetic stimulation on neonatal development in the premature infant. Dev. Psychobiol. 9, 569-578.

Wiesel, T.N., and Hubel, D. 1963. Effects of visual deprivation on morphology and physiology of cells in the cats lateral geniculate body. J. Neurophysiol. 26, 978-993.

Wiesel, T.N., and Hubel, D. 1965. Comparison of the effects of unilateral and bilateral eye closure on cortical unit responses in kittens. J. Neurophysiol. 28, 1029-1040.

Wilber, K. 1977. The Spectrum of Consciousness. Theosophical Publishing House, Wheaton, Illinois.

Will, B., and Rosenzweig, M. 1976. Effects de l'environment sur la recuperation functionelle apres lesions cerebrals chez les rats adults. Biol. Behav. 1, 5-16.

Will, B., Rosenzweig, M.R., and Bennett, E.L. 1976. Effects of differrental environments on recovery from neonatal brain lesions, measured by problem solving scores and brain dimensions. Physiol. Behav. 16, 603-611.

Will, B.E., Rosenzweig, M., Hubert, M., and Morimoto, H. 1977. Relatively brief environmental enrichment aids recovery of learning capacity and alters brain measures after postweaning brain lesions in rats. J. Comp. Physiol. Psychol. 91, 33-50.

Williams, M., and Scarr, S. 1971. Effects of short term intervention on performance in low birth weight disadvantaged children. Pediatrics 47, 289.

Wilson, J.R., and Sherman, S.M. 1975. Effects of monocular deprivation on cat striate cortex. Soc. Neurosci. Abstr. 1, 86.

Wilson, T. 1970. Normative and interpretive paradigms in sociology, in Understanding Everyday Life, Douglas, J. ed. Aldine, Chicago, Illinois, 57-79.

Wilson, M., Warren, J.M., and Abbot, L. 1965. Infantile stimulation, activity, and learning by cats. Child Dev. 36, 843-853.

Wing, J., Laff, J., and Hirsch, S. 1972. Preventive treatment of schizophrenia: Some theoretical and methodological issues. Paper presented at the meeting of the American Psychopathological Association, New York, March.

Wolf, A., and Von Haxthausen, E. Frhr. 1960. Aur Analyse der Wirkung einiger zentralsedativer Substanzen. Arzneim.-Forsch. 10, 50-52.

Wooley, D.E., Timiras, P., and Woodbury, D.M. 1960. Some effects of sex steroids on brain excitability and metabolism. Proc. West. Pharmacol. Soc. 3, 11-23.

Worsley, A., Cummins, R.A., Budtz-Olsen, O.E., and Walsh, R.N. 1973. Testosterone, early experience and emotional arousal in chicks. Proc. Aust. Soc. Med. Res. 3, 174-175.

Worthington, C.S., and Isaac, W. 1967. Occipital ablation and retention of a visual conditioned avoidance response in the the rat. Psychol. Sci. 8, 289-290.

Yanai, J., and Ginsburg, B.E. 1976. Increased sensitivity to chronic ethanol in isolated mice. Psychopharmacologia 46, 185-189.

Yarrow, L.S., Rubenstein, J.L., Peterson, F.A., and Jankowski, J.J. 1972. Dimensions of early stimulation and their differential effects on infant development. Merrill-Palmer. 18, 205-218.

Yen, H.C.Y., Stanger, L., and Millman, N. 1959. Ataractic suppression of isolation-induced aggressive behavior. Arch. Int. Pharmacodyn. 123, 179-185.

Young, P.T. 1959. The role of affective processes in learning and motivation. Psychol. Rev. 66, 109-125.

Zablocka, T., Zennicki, B., and Kosmel, A. 1976. Visual cortex role in object discrimination in cats deprived of pattern vision from birth. Acta. Neurobiol. Exp. 36, 157-168.

Zagorylko, T.M., and Sollertinskaya, T.N. 1958. Towards a comparative analysis of the mechanisms of orientation reactions, in The Orientation Reaction and Orienting-Investigating Activity. Academy of Pedagogical Sci. Moscow.

Zetterberg, A. 1966. Synthesis and accumulation of nuclear and cytoplasmic proteins during interphase in mouse fibroblast *in vitro*. Exp. Cell. Res. 42, 500-511.

Zigmond, M.H., and Harvey, J.A. 1970. Resistance to central norepinephrine depletion and decreased mortality in rats chronically exposed to electric foot shock. J. Neuro-Visc. Relat. 31, 373-381.

Zolman, J.F., and Morimoto, H. 1962. Effects of age of training on cholinesterase activity in the brains of maze bright rats. J. Comp. Physiol. Psychol. 55, 794.

Zolman, J.F., and Morimoto, H. 1965. Cerebral changes related to duration of environmental complexity and locomotor activity. J. Comp. Physiol. Psychol. 60, 382-387.

Zubeck, J.P. 1969. Physiological and biochemical effects, in Sensory Deprivation: Fifteen Years of Research, Zubeck, J.P. ed. Appleton-Century-Crofts, New York.

Zuckerman, M. 1969. Theoretical formulations, in Sensory Deprivation: Fifteen Years of Research, Zubeck, J.P. ed. Appleton-Century-Crofts, New York.

# Index

GPSR Compliance

*The European Union's (EU) General Product Safety Regulation (GPSR) is a set of rules that requires consumer products to be safe and our obligations to ensure this.*

*If you have any concerns about our products, you can contact us on ProductSafety@springernature.com*

In case Publisher is established outside the EU, the EU authorized representative is:

Springer Nature Customer Service Center GmbH
Europaplatz 3
69115 Heidelberg, Germany

**Batch number: 10584442**

Printed by Printforce, the Netherlands